AFRICAN AMERICAN WOMEN CHEMISTS

AFRICAN AMERICAN WOMEN CHEMISTS

JEANNETTE E. BROWN

Oxford University Press, Inc., publishes works that further
Oxford University's objective of excellence
in research, scholarship, and education.

Oxford New York
Auckland Cape Town Dar es Salaam Hong Kong Karachi
Kuala Lumpur Madrid Melbourne Mexico City Nairobi
New Delhi Shanghai Taipei Toronto

With offices in
Argentina Austria Brazil Chile Czech Republic France Greece
Guatemala Hungary Italy Japan Poland Portugal Singapore
South Korea Switzerland Thailand Turkey Ukraine Vietnam

Published by Oxford University Press, Inc.
198 Madison Avenue, New York, New York 10016

www.oup.com

Library of Congress Cataloging-in-Publication Data

Brown, Jeannette E. (Jeannette Elizabeth), 1934-
African American women chemists / Jeannette E. Brown.
p. cm.
Summary: "Beginning with Dr. Marie Maynard Daly, the first African American woman to receive a PhD in chemistry in the United States—in 1947, from Columbia University—this well researched and fascinating book celebrates the lives and history of African American women chemists. Written by Jeannette Brown, an African American chemist herself, the book profiles the lives of numerous women, ranging from the earliest pioneers up until the late 1960's when the Civil Rights Acts sparked greater career opportunities. Brown examines each woman's motivation to pursue chemistry, describes their struggles to obtain an education and their efforts to succeed in a field in which there were few African American men, much less African American women, and details their often quite significant accomplishments. The book looks at chemists in academia, industry, and government, as well as chemical engineers, whose career path is very different from that of the tradition chemist, and it concludes with a chapter on the future of African American women chemists, which will be of interest to all women interested in a career in science"—Provided by publisher.

Includes bibliographical references and index.
ISBN 978-0-19-974288-2 (hardback)
1. African American women chemists—Biography. 2. Chemists—United States—Biography.
I. Title.
QD21.B69 2011
540.92'2—dc23

2011013662

1 3 5 7 9 8 6 4 2
Printed in the United States of America
on acid-free paper

This book is dedicated to the memory of my mother and father, Freddie and Ada Brown, who worked so hard so their only child could realize her dream of being a chemist.

CONTENTS

ACKNOWLEDGMENTS

So many people helped make this book possible. I did not do it alone. Of course, I need to thank all the living women in the book whom I either personally interviewed, or asked to review what I had written about them. Most of the living women in this book have set the record straight so that the story that I tell is their story. Thank you for being so helpful and complying with my deadlines as the project neared its end. I hope you are pleased with the results.

I especially want to acknowledge that it was Dr. Marie Maynard Daly, the first African American woman chemist I ever met, who inspired me to study the history of other African American women chemists.

Then there are my personal editors who took time to read early drafts of the book, Hilary Domush, Shannen Dee Williams, Joan Fry, Doreen Blanc-Rockstrom, Olinda Young, and Nancy Danch. I want to especially thank Olinda Young, who read and edited the entire book, and Nancy Danch, who was there for me during the last hectic days. I also want to thank Margaret Wu, who found all the publications for these women using Sci Finder®. Thank you for taking time from your busy day to help me.

Thanks also to my cheerleaders Sandra Allen and Linda Buczynski who kept me on task researching, writing and editing the book. You are good friends.

I would also like to thank the many librarians and archivists who helped me wade through the references or lugged the boxes of information for me to see. They were also there online when I just needed a quick answer to a question. I don't know what I would have done without them.

Thanks to the members of the Women Chemist Committee (WCC) and the Committee on Minority Affairs (CMA) of the American Chemical Society (ACS) for their support and encouragement. And to those members who read the book proposal, thank you. I would especially like to thank Janet Bryant, Amber Hinkle, and Judy Cohen, who were the chairs of WCC during this process, and Allison Williams, chair of CMA. Thanks also to the members of the ACS Board of Directors for being supportive of this project, and also to the members of the North Jersey Section of the ACS for bearing with me during this process as I deferred my duties to work on the book. I will be back!

I want to thank the people who first encouraged me to write this book: Dr. Willie Pearson, and Dr. Ronald Mickens. Thank you for keeping me going, reminding me of the need for this book, and suggesting future projects. Special thanks go to Dr. Winni Warren, whom I have never met, but who encouraged me to keep going during our telephone conversations. I thank her for her support even though she was very ill at the time.

Thanks also to the members of the Princeton Research Forum (PRF) for allowing a novice writer to join their organization of Independent Scholars. I should also like to thank the members of PRF for encouragement, especially Dr. Karen Reed, chair of the science group.

Thank you to Dr. Henry Louis Gates, Jr., who became my mentor for this book even though he did not know it. He encouraged me to send my proposal to his publisher, Oxford University Press, because I had written about some of the women in *African American National Biography*. I want to thank him very much for his e-mails of encouragement. Another remote mentor is Dr. Deborah Gray White, Professor of African American History at Rutgers University, who encouraged two of her students to help me with the completion of this book.

Thank you also to the Chemical Heritage Foundation (CHF) and the sponsors of the two fellowships, which I obtained at CHF, the Société de Chimie Industrielle (American Section), and the Ullyot Fellowship. I am so sorry that Barbara Ullyott, the sponsor of the Ullyot Fellowship, did not live to see my book. Thank you also to the staff of CHF for their hospitality and friendship during my two fellowships.

Lastly I want to thank my editors at Oxford University Press, Jeremy Lewis and Hallie Stebbins, who worked so hard to get this book published while shepherding a novice author who was learning on the job. Thank you for helping me to realize my lifelong dream.

As I put the finishing touches to this book, I realize that I have only scratched the surface. There are so many African American women chemists whose work in laboratories or classroom has gone unsung and unnoticed.

Each has contributed in their own unique way to the advancement of the science of chemistry. Bravo! Keep up the good work.

Jeannette Elizabeth Brown
Hillsborough, New Jersey
March 24, 2011

AFRICAN AMERICAN WOMEN CHEMISTS

CHAPTER 1

The Reason for This Book and Why These Women Were Chosen

Many people have studied the history of African American women chemists, but the information is scattered in many references, articles, and trade books. Until now, there was no one place where one could access extensive information about these women. This book is a compilation of all the references to date about the lives of these women; the chapters include a brief biography of each woman, with citations to the published information. The back matter provides a list of references. Not all of the women that I have written about are primarily researchers; some of them chose to be educators or businesspeople. My selection includes women pioneers—women who were the first to enter the field and receive a degree in chemistry, biochemistry, or chemical engineering. Some of these women were able to work as chemists before obtaining an advanced degree in chemistry. They later chose to pursue the PhD degree when major colleges and university allowed all students, regardless of race, to study.[1] Some of the women chose not to pursue PhD degrees, ending their education with an MS degree. I extended my research to try to find the earliest women to pursue chemistry after the Civil War. It was difficult to find such early documents; however, I have not stopped searching.

The first woman in this book, Josephine Silone Yates, was born into a family of free blacks in the north in 1852, before the Civil War. The next woman, Bebee Steven Lynk, was born in Mason, Tennessee in 1872 but not much is known about her early life. Alice Ball was born in 1896 into a family of free blacks in Seattle. These women, who were born in the nineteenth century, studied chemistry. Only one obtained an advanced degree: a PhC, which may have been a two-year degree. Josephine Silone Yates is reputed to have obtained a master's degree.[2]

Most of the women in this book were, as the expression is used today, "nerds." They were outstanding students in school. Most were early readers; they learned to go to the library or to read books that their parents owned, and they did this even if their parents were not literate. They had other adults—teachers and members of the community—who were supportive of them. The women who lived in the South and went to segregated schools had teachers who were very supportive. Most of those teachers were black and had degrees not only in pedagogy but in the subject that they were teaching. This may be explained by the fact that black teachers who majored in science could not find employment in careers other than teaching, either in primary school or college, this is discussed in the history resources that are listed in Chapter 2. Two or more of the women in this book started their scientific careers as high school teachers before they could obtain other employment.

The choice of undergraduate college was automatic for some of these women. They did not really have a choice, since they could only go to a historical black college if they lived in the South before the era of the civil rights laws and equal access. As for graduate school, most attended the college suggested by their chemistry professor or other mentor. This may or may not have been another historical black college. In some cases they stayed on at their undergraduate institutions to obtain an MS degree. Most of them went to another university for their PhD; these included the University of Chicago, Iowa State, University of Iowa, Massachusetts Institute of Technology, Columbia University, Syracuse University, and Yale University, to mention a few. In many of those universities they were the first black women to receive a PhD from that university; Columbia University has the distinction of producing the first black woman PhD overall.

Twenty-two of the women in this book married, and of those, eight were divorced, some twice; the rest never married at all, choosing to remain single and to devote their lives to their careers and the mentoring of other young women. Of the women who married, there were five husband-and-wife teams in which both were scientists and worked together. One woman, Dr. Reatha Clark King, and her husband, arranged their careers so that they did not interfere with each other's chance for advancement. They followed each other and shared child care. Another woman, Dr. Esther Hopkins, was able to advance her career while still being a wife and mother. She and her son went to graduate school together; he was placed in child care while she studied. They were called the "odd couple." Of the eight women who were divorced, some had children, but motherhood did not interfere with their career advancement. I did not pursue the reasons for this, since that was not within the scope of this book.

I began this study of the lives of African American women chemists after meeting Dr. Marie Daly at a scientific meeting in 1984. Dr. Marie Daly is the first African American woman to receive the PhD in chemistry. She was very modest and unassuming. Since I am also a chemist, I decided to learn about the other African American women who decided to pursue a degree in chemistry in spite of the odds against them. As I began the research, I began to give talks about the lives of the women at scientific meetings. At many of those meetings, people asked me to write this book.

The women whom I chose for this book received degrees before the civil rights legislation changed the prospect for careers for all African Americans. I have selected only two women who were born in the 1950s, who would have benefited from the equal employment rules of the civil rights era. They were chosen because of their prominence. One, Dr. Cheryl Shepherd, was Undersecretary of Commerce in the administration of President Bill Clinton, and the other, Dr. Lynda Jordan, was featured in a PBS documentary about women in science.[3]

Most of the women in this book obtained advanced degrees, either a master's degree or a doctoral degree. Two women pursued law degrees and two are chemical engineers. However, unlike most women chemists, they did not begin to study for the doctoral degree immediately obtaining a BS degree, as most people do now. For most of the women, there were financial reasons for that. They needed to work in order to get the money for graduate study. (This may still be happening to women now; they may have to delay studying for a PhD.) For the women who came from the South and received their undergraduate degrees from the historically black colleges, they were able to become faculty members at those institutions with just a BS or MS degree. Most of the colleges in the North would not admit black scientists, either male or female, for an advanced degree in the early years of the twentieth century. One of the main exceptions was the University of Chicago, where three women went to receive their PhD degrees.[4] In addition, it may have been that since men who were studying chemistry could not receive advanced degrees from mainstream colleges and universities, the women did not try. Some of the men, like Dr. Percy Julian, went to Europe to receive an advanced degree.[5] I have not yet found women who may have done this.

In spite of these limitations, the accomplishments of these women is remarkable. It is even more remarkable because it has been hidden. No one knew the work of these women, or maybe others took the credit for their work.[6] I hope that the reader will enjoy reading these stories as much as I have enjoyed researching and writing them. Most of the living women in this book have written their own stories in order to set the record straight. I hope that these stories will inspire some readers to delve deeper into the lives of these women, or to research the lives of women whom I have not written about. I also hope that they will inspire other readers to enter the field of chemistry as a professor, teacher, industrial or government researcher, or as a student of chemistry as a background for other fields.

There are numerous careers for which such background knowledge is useful, such as law or scientific writing, to mention two.

I also hope that historians of African American women, of women and or of women in science will be interested in using this book as a springboard for further research.

CHAPTER 2

Resources for Historical Background

Many historians have written about the history of African Americans in science, but most of the articles focus only on the men and very little is written about the women. It would take additional research to find information pertaining only to the women. However, since both men and women lived through the same era, much of what affected the men also affected the women. The background information about black women chemists could probably fit into another book or at least a paper, but that was not within the scope of this book.

Dr. Wini Warren, author of *Black Women Scientists in the United States,* did some extensive research on the background history of black women in science, which she planned to put into a future book; due to health problems it was never written.[1] However, the Introduction to Dr. Warren's book is well worth reading for some of the background history of the women. The endnotes in that chapter provide an extensive bibliography about the history of blacks in science.[2] In addition, Dr. Warren includes an extensive discussion about the background history of black women scientists in the introduction of her thesis,

"Hearts and Minds: Black Women Scientists in the United States 1900–1960."[3]

Sisters in Science by Diann Jordan[4] features author interviews of black women scientists, some of whom are chemists. The Introduction of her book, discusses the background history. Dr. Jordan also includes a history of black colleges in the section "The Role of the Black College in Educating African American Scientists."[5] Since many of the women in this book had their first college education in a black college, it is worth reading.

Information about several of the African American women chemists in this book can be found in *Contributions of Black Women to America,* Volume 2.[6] The Introduction and Chapter 1 in the "Science" section give some background information about the history of women in science.

Dr. Willie Pearson Jr., professor of sociology at the School of History, Technology, and Society, Georgia Institute of Technology, has written several books and articles about African American scientists and chemists.[7] In his book *Black Scientists, White Society, and Colorless Science: A Study of Universalism in American Science,* Chapter 7, entitled "On Being Black Female and Scientist," discusses female scientists. For his book *Beyond Small Numbers: Voices of African American PhD Chemists,* he interviewed PhD chemists, some of them women, about their life and career choice. Chapter 1, entitled "The African American Presence in the American Chemistry Community: A Brief History," gives the background history of African American chemists.

For general background reading about African Americans in science, there are scholars who specialize in this subject. One of them is Dr. Kenneth R. Manning, who is currently the Thomas Meloy Professor of Rhetoric

(Program in Writing and Humanistic Studies and STS) at the Massachusetts Institute of Technology. He has written several articles about African Americans in science.[8]

Two other books worth reading for background information are *The Negro in Science*[9] and *Scientists in the Black Perspective.*[10] These two books provide a different perspective and also discuss some of the women in this book. *The Negro in Science* was written in the 1950s, before the Civil Rights Act; *Scientists in the Black Perspective* was written in 1974, after the Civil Rights Act. They both discuss the history of blacks in science, mostly PhD scientists and mostly men, and urge young black people to consider a degree in science as an option. This is especially true of the second book, since the opportunities had been opened to all races and students were encouraged to enter the field.

The books and articles mentioned above are not an exhaustive search of the information about the history of African Americans in science. However, it will give the reader a starting place for further research on the subject.

At the end of this book is a time line in which some of the key historical happenings are highlighted. The birth dates and, in some cases, the dates of death for the women in the book are inserted in the time line, which will give the reader some idea of the period of history in which the women lived. Because they lived in that period of time, the events of the era may have affected their lives.

CHAPTER 3

Early Pioneers

JOSEPHINE SILONE YATES

Born into a free black family in the early nineteenth century, Josephine Silone Yates was a pioneering woman faculty member at the historically black Lincoln Institute (now University) in Jefferson City, Missouri, where she headed the Department of Natural Sciences. Yates later rose to prominence in the black women's club movement of the late nineteenth and early twentieth centuries, serving as president of the famed National Association of Colored Women (NACW) from 1901 to 1905.

Josephine was born in 1852 in Mattituck, New York, to Alexander and Parthenia Reeve Silone. She was their second daughter. Her maternal grandfather, Lymas Reeves, had been a slave in Suffolk County, Long Island, New York, but was freed in 1813. Lymas owned a house in Mattituck, and Josephine's parents lived with him.[1] Josephine's mother was well educated for the time, and she taught her daughter to read and write at home. Josephine's earliest and fondest memories were of being taught to read from the Bible while snuggled on her mother's lap. Her mother made her call out the words as she pointed to them.[2] Josephine began school at age six, where her teachers immediately recognized her preparedness and advanced her rapidly through the elementary grades. At the age of nine, she reportedly studied physiology and physics and possessed advanced mathematical ability.[3] Silone also advanced her writing career at the age of nine, by submitting "a story for publication to a New York weekly magazine. Though the article was rejected for publication, she received a letter of encouragement, which increased her ambition to succeed."[4]

Josephine's uncle, Reverend John Bunyan Reeve, was the pastor of the Lombard Street Central Church in Philadelphia. Because of his interest in the education of his niece, he convinced his sister, Parthenia, to send Josephine at the age of eleven to live with him in Philadelphia so that

FIGURE 3.1 Josephine Silone Yates Credit: Photo courtesy of the Library of Congress

she could attend the Institute for Colored Youth directed by Fanny Jackson-Coppin. It was probably felt that Josephine's education would progress better under the mentorship of Jackson-Coppin.[5] The Society of Friends founded the Institute for Colored Youth in 1837. It served as a classical high school and included a preparatory department, a teacher training course, boys and girls high school department, and later, vocational courses.

With Jackson-Coppin as a role model, it was no wonder that Josephine, herself, became a leader and pioneer. Josephine was only able to stay at the Institute for a year because her uncle, Reverend J. B. Reeve, took a position in the Theology Department of Howard University. In order to continue her education, Josephine was sent to live with her maternal aunt, Francis I. Girard, of Newport, Rhode Island. At the age of fourteen she entered the highest grade of the grammar school there. Jackson-Coppin continued to mentor her from a distance, calling her "a brilliant example of what a girl may do."[6] Remember, this was a time when women spent most of their time doing housework and other domestic duties.

Josephine was the only black student in both her grammar school and, later, in Rogers High School in Newport; nevertheless, she excelled in her studies. In high school her science teacher considered her his brightest pupil. She showed a strong interest in chemistry and did additional laboratory work under her teacher's guidance. She graduated from high school in 1877, after only three years, became the valedictorian of her class, and also received a medal for scholarship. The first graduate of color at that high school, she was treated with respect by all her teachers and members of the school board. After high school, she took the examination to become a teacher and achieved the highest score recorded in Newport to that date.

Instead of going to a university to study, as her teachers urged her to do, Josephine decided to attend Rhode Island State Normal School to prepare for a career in teaching. This may have been due to the fact that her mentor Jackson-Coppin had completed the Normal School program in Rhode Island in 1860.[7] In 1879 Josephine graduated from the Rhode Island State Normal School with honors, the only black student in her class.

As is apparent, Josephine's early life and education differed from the majority of blacks living in the Southern and border states during the first generation after the Civil War. There is no indication that she had ever traveled to a former slave state. Therefore, it is interesting that her first offer of a position came from Lincoln Institute in 1881.[8] She was only twenty-three years old and arrived at the school fifteen years after it had been established as a black subscription school, eleven years after it was designated as Missouri's black normal school, and two years after the state took over its operation.[9] Josephine Silone was one of the first black teachers hired, first to teach chemistry, occasionally other natural sciences, and then to head the entire natural science department. She thus became the first black woman to head a college science department. She was also the first woman to be promoted to full professorship.[10] Few details are known about Silone's life as a teacher at Lincoln Institute. During this time Lincoln was a highly regimented and patriarchal institution. President Inman E Page kept the management of the school directly under his control. Faculty members lived on campus, as did virtually all the students; most of whom were drawn from the working class. Faculty members lived in the same dorms as the students, which meant that they could have direct control over their students.

Silone's attitude toward teaching was presented in a 1904 essay entitled "The Equipment of a Teacher," which was published under her married name, Yates. She said, "The aim of all true education is to give to body and soul all the beauty, strength, and perfection of which they are capable, to fit the individual for complete living."[11] Realizing that the audience for her essay may not be entirely African American, she assumed "that the primary task of the teacher was to encourage and teach behavior that would help to diminish negative racial stereotypes."[12]

She was aware of the charge that black teachers were too often unprepared or less prepared than whites. She urged them to be prepared for their responsibilities as teachers. Teaching, she wrote, "must be raised to the dignity of a profession and it must call forth dedicated men and women willing to devote their lives to its cause." She goes on to say, "To fail as a teacher was more than a personal failure, it was a betrayal of the race." Booker T. Washington was attracted to her success as a teacher and offered her the position of "lady-principal" of Tuskegee Institute if she would leave Lincoln. However, she declined the offer.

In 1889, Josephine Silone likely had to resign from her position at Lincoln Institute in order to get married to William W. Yates. This is surprising in light of her criticism of women who taught only until a suitable suitor came along. At the end of the nineteenth century, married women were forbidden to teach in many communities. Her fiancé was the principal of Phillips School in Kansas City, Missouri; therefore, she moved from Jefferson City to Kansas City in order to get married. She had probably met Yates in 1885, when the two of them appear on the same program at an annual meeting of the state teachers' association.

Her activities during the first part of her marriage are unknown, but she did have two children, a daughter, Josephine Silone Yates, Jr., born in 1890, and a son, William Blyden Yates, born in 1895. She probably taught in the Kansas City black school and tutored students who could not attend public school. She was also a writer on a variety of subjects, mainly for newspaper publications.

Josephine Silone Yates needed another outlet for her intellectual skills, as did many married middle-class women of the time, both black and white, so she turned to the Women's Club Movement. In 1893 she started and became first president of the Women's League of Kansas City, and was active in other organizations for African American women. The first project of the Women's League was the establishment of an industrial home and school for teaching cooking, sewing, and other useful employments. They started a kindergarten and purchased a house that became a "Home for Working Girls."

Very much involved in the women's movement, Yates became the correspondent for the *Woman's Era*, a monthly magazine published by black women in the United States. The black women's clubs joined together to become the National Association of Colored Women (NACW); Yates later became the fourth vice president of that organization. In 1899, Yates presented an outstanding paper at the organization's biennial meeting in Chicago. W. E. B. DuBois praised Yates as "perhaps the finest specimen of Negro womanhood present. . . ." Her paper was entitled "An Equal Moral Standard for Men and Women." Yates was elected president of the NACW in 1901. This gave her an opportunity to speak and write about the status of black women and to offer advice about the way that they should present themselves.

Yates was asked to return to teach at Lincoln Institute for the 1902–1903 year. She became chairperson of the departments of English and history and advisor to women. In 1908 she wanted to resign, but the Regents of the university refused her resignation.

In 1910 her husband died and she returned to Kansas City to take care of her family. She worked for the Kansas City Board of Education, teaching young people at Lincoln High School. She died at age fifty-three in 1912 after a two-day illness.

Josephine Silone Yates was a writer under her name as well as the pseudonym R. K. Porter. She was the editor of the *Negro Educational Review* and wrote for the *Boston Herald*, *Transcript*, *Los Angeles Herald,* and the *Pacific*.

BEEBE STEVEN LYNK

Beebe Steven Lynk was a pioneer in chemistry, but little is known about her. She is one of the women in this book for whom more research could be done.

She was born in Mason, Tennessee, on October 24, 1872, the daughter of Henderson and Judiam (Byod) Steven. Little is known about her early life, her parents, or whether she had siblings. However, her parents must have been committed to the education of their daughter, because she attended Lane College[13] in Jackson, Tennessee, and graduated in 1892. Given that she would have been about twenty years old, it is not clear whether she attended a two-year college or started at age sixteen.

On April 13, 1893, she married Dr. Miles V. Lynk. Her career is linked to her husband's career. Dr. Miles Lynk received his MD degree from Meharry Medical School in 1891. He was the founder, editor, and publisher of *Medical and Surgical Observer* (1892), which was the first medical journal issued by a black man in the United States. In 1900, he founded the University of West Tennessee, a school for blacks.

Beebe Lynk studied pharmaceutical chemistry at the University of West Tennessee in 1901 and earned a PhC, which was a two-year degree, in 1903.[14] She then became the professor of medical Latin botany and *materia medica* at the university's new medical school.[15] A photo of the medical school faculty shows two women out of ten faculty members, including Mrs. Lynk.[16]

Like Josephine Silone Yates, Beebe Lynk was active in the women's club movement. She wrote a book entitled *Advice to Colored Women*, which was published in 1896. Not much more is known about her. The fact that she had a degree in chemistry and taught at the college level in the late nineteenth century is significant. More research on Beebe Lynk may be available via the Tennessee State archives, either online or in person, by searching the history of the black women's movement.

ALICE AUGUSTA BALL

Alice Ball was a pharmaceutical chemist who in 1915, at the age of twenty-three, developed a treatment for

Hansen's disease,[17] which continued to be the most effective treatment until the 1940s, long after her death.

Alice Augusta Ball was born in Seattle, Washington, on July 24, 1892, the third of four children. Her mother was Laura L. Howard Ball and her father was James Presley Ball, Jr., a lawyer and photographer. Her grandfather, James Presley Ball, was a well-known nineteenth-century photographer.[18] Her family lived in Seattle until 1902, when her grandfather's failing health due to rheumatism caused the family to relocate to Hawaii for an improved climate.

Their home and photography studio were situated in downtown Honolulu. Since Alice's mother was also a photographer, along with her father, grandfather, and aunt, Alice learned about the chemicals that were used to develop

FIGURE 3.2 Alice Augusta Ball, detail of Graduating Class photographs, GC 1915, in University Archives Credit: Photo courtesy of the University of Hawaii Library

film at that time.[19] Alice attended school at Central Grammar (now Central Intermediate) in Honolulu from 1902 to 1904. J. P. Ball, Sr., did not live long in Hawaii; he died on May 3, 1904. After his death, the family moved back to Seattle, where Alice entered Seattle High School in 1906, earning excellent grades, especially in the sciences.[20]

After graduating from high school, Ball attended the University of Washington, where she studied chemistry and received two BS degrees, one in pharmaceutical chemistry (1912) and another in pharmacy (1914). She also co-published with her pharmacy instructor, William M. Dehn, a ten-page article on "Benzoylations in Ether Solution" in the *Journal of the American Chemical Society.*[21] She had her choice of two scholarships to study for a master's degree, one from the University of California at Berkeley and the other from the College of Hawaii (now the University of Hawaii). "She chose the latter, perhaps because she was familiar with the Islands."[22] In Hawaii, she first stayed at the YWCA residence, the "Homestead," and then moved to the MacDonalds, which was a hotel for women. It was remarkable that she was admitted to the College of Hawaii as its first graduate student. In the years 1914 and 1915, men dominated higher education; therefore, she was a brave woman to return to Hawaii, where she had no family.[23]

At the College of Hawaii, Alice spent one year (1914–1915) on her studies for her MS degree, graduating on June 1, 1915. Her master's thesis involved the identification of the active constituents of Kava root. The title of her master's thesis is "The Chemical Constituents of Piper Methysticum; The Chemical Constituents of the Active Principle of the Ava Root."[24] She was the first woman and also the first African American to graduate with a master's degree from the college. Upon obtaining this degree, she became a unique woman for this time in the United States, because few African American women had received advanced degrees in chemistry.[25] She then taught chemistry at the College of Hawaii (1915–1916) as the first black instructor in the Chemistry Department.[26]

While she was teaching at the College of Hawaii, Dr. Harry T. Hollmann recruited Ball to work on research to make chaulmoogra oil more effective in treating patients with Hansen's disease. Hansen's disease, also called leprosy, was named after Gerhard Henrik Armauer Hansen, the Norwegian physician who discovered that leprosy is caused by two strains of bacteria, *Mycobacterium leprae* and *Mycobacterium lepromatosis*.[27] In the early twentieth century, leprosy was considered very contagious, and people with the disease were isolated in communities called leper colonies (there was one leper community in Hawaii called Kalaupapa). The best palliative cure for leprosy at the time was the topical application of chaulmoogra oil. Chaulmoogra oil was obtained from seeds of the chaulmoogra tree, or *Taraktogenos kurzu*, which grows in Hawaii. The results of applications at that time were inconsistent because there were two kinds of oil, the true oil and false oil. When chaulmoogra oil was applied topically to the affected areas, the patients suffering from Hansen's disease showed some improvement. The patients showed greater improvement when given the oil orally, although the bitter taste was nauseating. Therefore, scientists needed to research a more effective method of administering the medicine. They could not inject it, as oil it was insoluble in water and painful when injected. (The active components of chaulmoogra oil are chaulmoogric acid and hydrocarpic acid.)

Dr. Harry T. Hollmann enlisted Ball to work on the chaulmoogra oil problem. He was the assistant surgeon at Kalhi Hospital in Hawaii, where new Hansen's disease patients were brought. Dr. Hollmann had also been the doctor at the federally funded Leprosy Investigation Station at Kalaupapa in 1909. Dr. Hollmann chose Alice Ball, the new instructor in chemistry at Hawaii College, to work on the problem because of her expertise in laboratory work. Her task was to develop an effective treatment for Hansen's disease. She worked hard, running many experiments, and came up with a solution by simply preparing the ethyl esters of the two fatty acids; thereby, making them water soluble and injectable. This was something that many far more accomplished pharmacologists and chemists had not previously been able to do.[28] Thus, at age 23 she developed what became known as the "Ball method," for treating Hansen's disease.

She did her research at night while teaching her classes during the day. Apparently, soon after her discovery and during one class, she inhaled chlorine gas. With no ventilation hoods in her laboratory, she became gravely ill, and was unable to publish the results of her discovery. She left Hawaii in October 1916 to return to Seattle and died on December 31, 1916 at the age of 24.[29]

After her death, Dr. Arthur L. Dean who was a chemist and president of the College of Hawaii continued Ball's research. Large quantities of this new drug were made and distributed to patients worldwide.[30] Dean published his results without mentioning the work of Ball and it became known as the Dean method. Later in a medical journal publication in 1922, Hollemann mentioned the contribution of Alice Ball.[31] Still it took years before Alice Ball was recognized for her accomplishments. The neglect may have been due to both sexism and racism which may explain why both birth and death certificates list her and her parents as white . . . this may have made life easier for them.[32]

In 1917, College of Hawaii students and faculty passed a resolution so Ball could "be an example to all her companions and associates at the College."[33] In 1990, a group of scholars, scientists, and residents of Kalaunana, gave homage to Ball for her outstanding work. Ball's accomplishments have survived in the historical record because of two individuals, Dr. Kathryn Takara of the University of Hawaii, who in 1977 began to research black women in Hawaii and discovered Alice's story, and Stanley Ali, a retired federal worker who was researching blacks in Hawaii. It was Ali who was able, with support from the University of Hawaii faculty, to have a portrait of Alice Ball hung in the Hamilton Library on campus. On February 29, 2008, the University of Hawaii honored Ball with a plaque placed by the chaulmoogra tree that is on campus. Lieutenant Governor Mazie Hirono of Hawaii named February 29 "Alice Ball Day," which is now celebrated every four years in Hawaii. In 2007, Alice Ball was posthumously awarded the University of Hawaii's Regents' Medal of Distinction.[34]

ESLANDA GOODE ROBESON

Eslanda Goode Robeson, wife of the singer and actor Paul Robeson, began her working career as a chemist before her marriage. She also became an anthropologist and activist.

Eslanda, or Essie, as she was called, was one of three children born to Eslanda Cardozo Goode and John Goode.

She was born on December 15, 1896, in Washington, D.C. She was a member of a middle-class family, since her father John Goode was a clerk in the War Department and her mother was a beauty culturist and osteopath. Her mother was from a family of free blacks in South Carolina. Essie's grandfather was Francis Louis Cardozo, a graduate of the University of Glasgow, and pastor of the Congregational Church in New Haven, Connecticut. Her grandfather obtained a grant from the American Missionary Association to found the Avery Institute, the first black secondary school in Charleston, South Carolina. He later held the post of state treasurer during the Reconstruction period after the Civil War.

Essie grew up in segregated Washington, D.C. When she was six years old, her father died and her mother chose to move the family to New York City so that her children could go to school in an integrated school system.[35] It was in New York that the elder Eslanda studied osteopathy and opened her own practice, where she had wealthy clients like Mrs. Joseph Pulitzer. In 1912, the family moved to Chicago because Essie's mother took over a beauty shop in that town.[36]

Essie was a bright youngster and graduated from high school in 1914 after only three years. She placed third in a statewide competitive exam and won a scholarship to attend the University of Illinois. There she registered to major in domestic science, probably because it was unusual for a black woman to major in chemistry at that time. She later discovered that it was the science part of domestic science that she liked. She was bored with her courses after two years, left the University of Illinois, and went to Teachers College of Columbia University to study chemistry.

She earned a BS in Analytical Chemistry from Columbia in 1920.[37] She also attended Columbia's medical school for one year. Since in 1918 the United States was involved in World War I and there was a shortage of scientists, Essie was encouraged by her advisor to apply for a position as a chemist at Columbia's Presbyterian Hospital. She became the first black person to obtain employment as an analytical chemist and technician in the surgery and pathology department at Columbia Presbyterian Medical Center. She was in charge of the laboratory, where she engaged in several research projects.[38]

In the autumn of 1920, Essie first met Paul Robeson at Devann's Restaurant in Harlem. He was already a well-known person, having graduated Phi Beta Kappa from Rutgers University. He was an outstanding athlete at Rutgers in football and basketball and had moved to Harlem in 1919 to study law at Columba University. Essie met him again when he was brought into Presbyterian Hospital with a football injury. It was love at first sight for Essie, but not so for Paul. Essie set her sights for him and was successful. They were married on August 17, 1921, in Port Chester, New York.

Paul Robeson graduated from Columbia Law School with honors and obtained a position at a prestigious white law firm. The assignments he was given did not challenge him, and he realized that being a black lawyer in that period of time would not be special. Instead, Essie encouraged Paul to get involved with the Players group at the Harlem YMCA because of his outstanding baritone voice; he was also working with the Provincetown Players. He became a success and was offered the lead in a play in London, which Essie encouraged him to take. Because of his successes, he dropped his law career and, in 1925, Essie resigned her position at Columbia Presbyterian Hospital to become his manager. In 1927, Essie and Paul celebrated the birth of their only child, Paul Robeson, Jr. (Paulie). Essie continued to act as Paul Robeson's manager for the rest of her life, arranging his performances in plays such as *Emperor Jones* and *Showboat*, and she traveled with him on tours to the cities of Europe and the Soviet Union.[39] Although Paul was becoming a stage success performing in white theaters and concert halls in Europe, he encountered racial prejudice in the United States.

In the late 1930s, after an extended trip through Africa, Essie Robeson began to study anthropology, first at the University of London, then at the Hartford Seminary Foundation. She was awarded a PhD in anthropology from Hartford in 1945. That year she published *African Journey*, a book that documented her travels in Africa.[40]

The rest of her career Essie Robeson revolved around Paul Robeson's career. Eslanda Robeson died of cancer in Beth Israel Hospital in New York on December 13, 1965.

ANGIE TURNER KING

Dr. Angie Turner King was an academically trained chemist and mathematician, who mentored many future scientists.

She was born Angie Lena Turner in Elkhorn, West Virginia, on December 9, 1905. Elkhorn was a racially segregated coal mining community in MacDowell County, West Virginia. She was the grandchild of Virginia slaves who were given land, a steer, and a log cabin upon emancipation.[41] Her mother died when she was eight years old, and not long after, her father was killed in a coal mining accident.[42] She was sent to live with her maternal grandmother, who verbally abused her because she was dark-skinned (her grandmother was very light-skinned and discriminated against dark-skinned African Americans). Eventually, she went to live with her grandfather, probably in Elkhorn. Even though her grandfather was illiterate, he insisted that she go to school. She graduated from high school in 1919 at the age of fourteen, with a good academic record, and was encouraged by her teachers to attend college.

She attended Bluefield Colored Institute for Teacher Training[43] for a number of years, and then matriculated at West Virginia State College. She worked to finance her tuition by waiting on tables and doing other odd jobs. She graduated from the West Virginia State, cum laude, in 1927 with a BS degree in chemistry and mathematics. She then began her teaching career at the laboratory high school at West Virginia State, where she taught for eight years. While teaching in the high school, she enrolled in Cornell University during the summers, to obtain an MS degree in both mathematics and chemistry. Her research-based master's thesis in chemistry was entitled "The Interaction of Solutions of Tannic Acid and Hydrous Ferric Oxide." Her thesis advisor for this work was probably T. R. Briggs. whom she acknowledges for his patient assistance.[44] She earned an MS degree from Cornell in 1931.

Upon her graduation from Cornell, she continued to teach at the laboratory school and then accepted a position at West Virginia State College. She is quoted as saying that her first order of business at the college was "to get a chemistry lab fixed up, so the students would know what a real laboratory was like."[45] During World War II, she taught as part of a program called the Army Specialized Training Program (ASTP) held at the college in 1943–1944. Angie Turner was one of the faculty members who taught the solders. She later recalled that it was hard to get the men seated in her chemistry class without formality since they were used to taking orders given by officers as to when to sit.[46]

She married Robert Elemore King on June 9, 1946, and had five daughters while working and studying. She continued her studies by enrolling in the University of Pittsburg to obtain a PhD in chemistry and mathematics. She again paid her graduate tuition herself and did not apply for financial aid. The title of her dissertation is, "An Analysis of Early Algebra Textbooks Used in American Schools before 1900."[47] She received her PhD in Mathematics in 1955. She was selected as West Virginia State College Alumnus of the Year in 1954.

Her husband died in 1958, but she continued to teach at West Virginia State, retiring in 1980. However, she continued to live in her house on campus. Dr. King died on February 28, 2004. She is known mostly for her teaching and mentoring of students who have gone on to receive advanced degrees.[48]

MARY ELLIOTT HILL

Mary Elliott Hill was a woman chemist who was both an excellent researcher and academic teacher. She worked on research instrumental in the development of plastics.

Born Mary Elliott on January 5, 1907, in the segregated South Carolina town of South Mills, she was the daughter of Robert Elliott and Frances Bass. She had two brothers. In 1925, she entered the historically black Virginia State College for Negros, later to be known as Virginia State University.[49] In 1929, she received a BS degree in chemistry from Virginia State University.

Mary met Carl McClellan Hill, the man who became her husband, when she was sixteen.[50] A classmate introduced them at a party. They were married two years later, during her sophomore year at Virginia State.

She began her career teaching chemistry at the Laboratory High School at Virginia State in 1930. From 1932 to 1936 she also taught college-level chemistry part-time at Hampton Institute, becoming a full-time faculty

FIGURE 3.3 Mary Elliott Hill Credit: © The Courier-Journal

member at the associate level in 1937. Until about 1950, racial barriers kept blacks from finding employment in the chemical industry or in academia.[51] There were few professional positions for African Americans in science, except within historically black colleges and universities. Even blacks with PhDs in science were forced to teach high school or to switch professions if there were no vacancies in the black colleges.[52]

Despite future employment hardship, Hill pursued an advanced degree during summers at the Graduate School of Arts and Science at the University of Pennsylvania and received an MS in analytical chemistry in 1941. She became an instructor at Dudley High School in Greensboro, North Carolina, after completing her degree. In 1944, she served for one year as assistant professor at Bennett College and then accepted a position at Tennessee A&I (Agricultural and Industrial), now Tennessee State University, where she was an associate professor of chemistry for eighteen years. In 1951, she became the acting head of the chemistry department. Her husband served as dean of the School of Arts and Sciences at Tennessee A&I. In 1962, her husband became president of Kentucky State College (KSU), also a historically black college. Fortunately, Mary Hill obtained an appointment, as professor of chemistry at KSU; sometimes a "trailing spouse" was not given the opportunity to obtain such a lofty position.[53]

Mary Hill and her husband worked as a research team. They were among the U.S. scientists who were fortunate enough to work in the period during World War II when the United States was breaking its ties to the German chemical industry and was beginning to initiate a chemical industry in the United States. Mary Hill was a brilliant woman who was able to succeed in a career in which at the time there were few female scientists, let alone black female scientists. The team of Mary and Carl Hill advanced the chemical industry by studying the Grignard reaction and making it work to produce ketenes, highly reactive chemical compounds with great potential as starting materials, for creating new types of ethers, and for helping a chemist to better understand complex chemical reactions. These relatively new chemicals, known as Grignard reagents (named after the inventor, Nobel laureate Victor Grignard), controlled rearrangements of atoms when reacting at a double bond like those found in ketenes. In her role as analytical chemist for her husband's research team, Hill created new analytical methodology and modified existing methods of ultraviolet spectroscopy. She established procedures for monitoring reaction progress by determining the degree of solubility of reaction species in non-aqueous reaction systems. This ability to isolate, identify, and quantify such reaction products enabled synthetic chemists on her team, aided by the Grignard reagents, to design new material, including plastics.[54] Hill was a coauthor on over forty research papers, though she was never the senior author. She also collaborated with her husband in the writing of two textbooks: *General College Chemistry* (1944), coauthored with Myron B. Towns; and *Experiments in Organic Chemistry* (1954), a laboratory manual that went through four editions.

Mary Hill preferred the classroom to the laboratory because she liked interacting with students. As a teacher, she demanded excellence from her students. She established student affiliate chapters of the American Chemical Society at the two campuses where she taught, possibly allowing these black schools to be the first to have such a group. This group influenced many African American students to consider careers in science and teaching. A conservative estimate is that at least twenty of her students became college professors.[55] The Manufacturing Chemists Association designated her as one of the top six chemistry teachers in the United States.

A reporter interviewed Mary Hill in 1963 about the growing number of women who were entering the field of chemistry. In explaining why it remained a challenge to lure women into the field, she commented, “Where laboratories have a choice between a man and a woman chemist, they will still choose a man nine times out of ten.” (Note: this was an explanation of the challenge of luring women into chemistry at the time.) She went on to say, “Today’s emphasis on science tends to attract girls, but the prestige value wanes once they realize the physical work, the self-discipline, and the study that chemistry requires.” She went on to say; “Girls like to be dressed up and feminine—high heels and long hair are not only out of place in a lab, they are unsafe. Don’t think this doesn’t influence a lot of prospects.”[56]

The Hills had one child, a daughter named Doris. Mary Hill was a working mother and wondered if that was good for her only child, who grew up to be teacher. The Hills loved to garden, which was the chief hobby of her husband. They were also very active in the church and community. She was active in the Presbyterian Church in both Nashville, Tennessee, and Frankfort, Kentucky. She was a member of the American Chemical Society, Tennessee Academy of Science, National Institute of Science, Alpha Kappa Alpha National Honor Society, and Beta Kappa Chi Sorority. She died on February 12, 1969.

CHAPTER 4

Marie Maynard Daly

Dr. Marie Maynard Daly was the first African American woman chemist to receive a PhD in chemistry. In addition, she was part of a research team that was working on the precursors to DNA.

Marie was born Marie Maynard Daly on April 16, 1921, to Ivan C. Daly and Helen Page, the first of three children. Her father, who had emigrated from the West Indies, received a scholarship from Cornell University to study chemistry; however, he had to drop out because he could not pay his room and board, and he became a postal worker. Daly's interest in science came from her father's encouragement and the desire to live his dream."[1] He later encouraged his daughter to pursue his dream, even though she was a woman and had brothers who were twins. In the 1920s, as a result of the women's suffrage movement, some women began to aspire to achievement in areas outside the domestic sphere. Marie's mother encouraged reading and spent many hours reading to her and her brothers. Marie's maternal grandfather had an extensive library, including books about scientists, such as *The Microbe Hunters* by Paul De Kruff; she read that book and many others like it.

Growing up in Queens, one of the boroughs of New York City, she attended the local public school, where she excelled. She was able to attend Hunter College High School, an all girls' school affiliated with Hunter College for women.[2] Since this was a laboratory school for Hunter College, the faculty encouraged the girls to excel in their studies. Since Marie had an aptitude for science, the teachers there encouraged her to study college-level chemistry while still in high school.

One of the many advantages of living in New York City during that time was that students who had good grades could enter one of the tuition-free colleges run by the City of New York. As a result, Daly enrolled in Queens College, then one of the newest institutions in the City College system, in Flushing, New York. Since Queens College was one of the newest colleges in the City College system, the classes were small. She graduated magna cum laude in 1942 with a BS degree in chemistry and she was designated as a Queens College Scholar.[3] Upon her graduation, the chemistry department at Queens College offered Daly a job as a part-time laboratory assistant. She also received a fellowship to study for a master's degree in chemistry at New York University, where she received her MS degree in 1943 during the height of World War II. Daly knew that she should continue her studies to receive a PhD degree. However, she needed money to pay her tuition, so she continued to work for a year at Queens College in order to save.

Due to the heightened need for scientists during World War II, Daly was able to receive a university fellowship from Columbia University to study for a PhD degree with Dr. Mary L. Caldwell. Dr. Caldwell's lab included many

FIGURE 4.1 Marie Maynard Daly Credit: Photo courtesy of Einstein College of Medicine Photographer: Ted Burrows

women as graduate students. Dr. Daly wrote in the obituary for Dr. Caldwell the following:

> [Dr. Caldwell] inspired her students with respect for technical excellence as well as fine scholarship. Her manners were rather formal; she rarely addressed students by first names and scrupulously changed the "Miss" or "Mr." to "Dr." immediately following a successful thesis defense. Despite her formal manner, she conveyed a sense of concern for a student's personal welfare. She could summon a bright word of encouragement when the work was not progressing fast enough; often ending her comments with a philosophical, "Well, child, that's research!"[4]

Dr. Caldwell is to be applauded for taking on an African American student at a time when there were few African

American students in chemistry. In 1947, Marie completed her research and in 1948, upon completion of her dissertation thesis, she received her PhD. She was the first African American woman to receive a doctorate in chemistry. The title of her PhD thesis is, "A Study of the Products Formed by the Action of Pancreatic Amylase on Corn Starch." This was a study of the mechanisms of action of pancreatic amylase, an enzyme, to break corn starch, a carbohydrate, into sugars, which are more available to human digestion. Dr. Daly dedicated her thesis to her mother and father and thanked Dr. Caldwell for suggesting the thesis project.

She wanted to begin her career with the research team headed by A. E. Mirsky at the Rockefeller Institute, but Dr. Mirsky was reported to say that she would need independent funding. Therefore, she applied for funding from the American Cancer Society and took a position at Howard University as an instructor of physical science. She was hired by the noted black physicist Herman R. Branson; in fact, she credits him with giving her her first job, which was a remarkable feat at the time.[5] Dr. Branson was one of the discoverers of the double helix, along with Francis Crick, but he was not mentioned in the Nobel Prize award.[6]

In 1948, she received a grant from the American Cancer Society and left Howard University for Rockefeller Institute, where she became the only African American scientist working there. The black community was excited about her work. There was an article published about her, entitled "Young Negro Scientist Engaged in Cancer Research." It states in part:

> "The youthful Doctor Daly is putting in her third year of research as a fellow of the American Cancer Society. Although her nose has been buried in cell nuclei, chromosomes,

> amylases and other bio-chemical, what-have-you, during the major part of her short adult life, Miss Daly is an all-around, life-loving American girl and is impervious to the attention she probably attracts as the only Negro scientist at the center".[7]

It is said that she enjoyed working at Rockefeller Institute. She had the opportunity to talk to legendary figures of science such as Leonor Michaelis and Francis Peyton Rous (Nobel laureate, 1966).[8] Her research was the study of the composition and metabolism of components of cell nuclei. This was the era in biochemistry when James Watson and Francis Crick were working on research that led to the structure of DNA. Dr. Daly worked on the research team with Alfred Ezra Mirsky and Vincent G. Allfrey that focused on cancer research. Her initial studies on the purine and pyrimidine content of DNA preceded the work by Watson and Crick by six or seven years. Cyril L. Moore, a former student, later noted, "To top this off, the basic amino acid content of histones and the properties of these important molecules were evaluated by her in the early fifties. Today we recognize the importance of these molecules in Chromatin structure and the functional organization of DNA."[9] The work of Allfrey, Daly, and Mirsky[10] was cited by Watson in his Nobel Prize address.[11]

Dr. Daly worked at the Rockefeller Institute for seven years before moving in 1955 to the College of Physicians and Surgeons at Columbia University, where she taught biochemistry. She also became a research associate at Goldwater Memorial Hospital, working with Dr. Quentin B. Deming, where she studied the underlying causes of heart attacks. Her early research involved the study of the metabolism of the arterial wall and how this process is related to aging, hypertension and arteriosclerosis. She discovered the role that cholesterol plays in heart problems and the effects of sugar and smoking on the heart.[12]

In 1960, Daly and Deming moved their research team to Yeshiva University at the Albert Einstein College of Medicine, where she became an assistant professor of biochemistry.[13] She was to work on the biochemistry of aging. Dr. Daly's grant-funded research focused on the breakdown of the circulatory system caused either by aging or hypertension. A letter of support for a Health Research Council grant by the City of New York, written by Dr. White of Albert Einstein College of Medicine, Department of Biochemistry, states that Dr. Daly exhibited a high degree of originality and industry in planning and carrying out the experiments. Dr. Daly planned a long-term study relating to the role of diet and endocrine factors in development of experimental hypertension.[14] Dr. Daly received the grant from the Health Research Council of New York in 1962 and held this position until 1972.

In 1971, she was promoted to associate professor, a position that she held until her retirement in 1986. In her last years at Einstein, her research focused on understanding the uptake of creatine by muscle cells by the in vitro synthesis of radio-labeled creatine.[15] At Einstein, Dr. Daly taught medical and graduate students. Dr. White stated that Dr. Daly was a very good teacher, especially with small groups. She also mentored many students at Einstein, one of whom was Dr. Francine Eissen, the first African American woman to receive a PhD in biology (now a retired professor of biology at Douglass College in New Jersey). Dr. Eissen credits Daly as her mentor.[16]

Dr. Daly was involved in the recruitment and training of minority students and guided the careers of African American students at the school. To that end, Dr. Daly was a member of the faculty committee that ran the Martin Luther King–Robert F. Kennedy Program that was established in 1968.[17] This was a program that prepared selected African American students for admission to Einstein. In a letter to the Dean of Albert Einstein College Dr. Walker, chair of the biochemistry department, credited her with developing the concept of the program, organizing the recruitment of students, developing the curriculum along with the committee, and teaching sections of the program. She became the administrator of the program and the cohesive force holding it together. She also was responsible for the cooperative efforts among the medical schools of New York in the recruiting and training of African American and Puerto Rican medical students.[18]

In 1988, Dr. Daly contributed a gift to Queens College to be used for scholarship aid to black students in the physical sciences. Aid coming from a black scientist, she felt, would provide a special motivation for black students.[19] Dr. Daly was very active in many professional societies. She was a member of the National Organization for the Professional Advancement of Black Chemists and Chemical Engineers (NOBCChE). On May 3, 1980, she was the moderator on the panel "Black Women in Chemistry, Biochemistry and Chemical Engineering: Confronting the Professional Challenges."[20] Dr. Daly's other professional memberships included membership in the American Chemical Society, New York Academy of Science, and American Association for the Advancement of Science Board. Further memberships included the Board of Governors, New York Academy of Science, Harvey Society, American Society of Biological Society, Fellow Council on Arteriosclerosis, American Heart Association, National Association for the Advancement of Colored People (NAACP), National Association of Negro Business and Professional Women, and Sigma Xi. She was also a fellow of the American Association for the Advancement of Science.

In 1985, she was nominated by the author for the American Chemical Society's Garvan Medal (now called the Olin Garvan Medal).[21] When the author contacted her about this, she did not think that she deserved to be nominated for this award because her thesis adviser, Letitia Caldwell, had won this award. The nominators were: Dr. Anne Brisco, assistant professor of Medicine College of Physicians and Surgeons Columbia University, Dr. Shirley Malcolm of the American Association, and Dr. Cyril L. Moore of Morehouse Medical School, a former student. Although the nomination was very strong, it was not successful.[22] In 1999, for her work as a researcher and professor of chemistry, Marie Daly was selected by the National Technical Association as one of the top fifty women in science, engineering, and technology.[23]

In 1961, Dr. Daly married Dr. Vincent Clark, who was a physician at Harlem Hospital in New York. They lived in Queens and later on Long Island. She was highly cultured and especially devoted to playing the flute. In later years, when cancer interfered with her ability to play the flute, Dr. Daly learned to play the guitar. She also was an excellent gardener and was devoted to her dogs.[24] She had no children of her own but had two stepchildren. She died on October 23, 2003.

CHAPTER 5

Chemical Educators

JOHNNIE HINES WATTS PROTHRO

Johnnie Hines Watts Prothro was one of the first African American women scientists and researchers in the field of food chemistry and nutrition. Having grown up in the segregated American South, Dr. Protho became particularly interested in promoting healthy nutrition and diets for African Americans.

Johnnie Hines Watts was born on February 28, 1922, in Atlanta, Georgia, in the segregated South. Her parents emphasized the importance of an education and she graduated from high school at the age of fifteen. She enrolled in the historically black Spelman College in Atlanta as a commuter student and received a BS degree with honors in Home Economics from Spelman in 1941. Following her graduation, she obtained a position as a teacher of foods and nutrition—the usual career path for African American women who earned bachelor's degrees in science during the Jim Crow era—at Atlanta's all-black Booker T. Washington High School. Watts taught at Booker T. Washington High School from 1941 to 1945, then moved to New York City to attend Columbia University, from which she received her MS degree in 1946.[1]

Armed with her master's degree, Watts became an instructor of chemistry at a historically black Southern University in Baton Rouge, Louisiana. She worked there during the 1946–1947 academic year before deciding to pursue a PhD. Watts enrolled in the University of Chicago after researching the doctoral offerings of several universities. She was the recipient of a number of scholarships and awards at the University of Chicago. Among the awards were the Laverne Noyes Scholarship (1948–1950), the Evaporated Milk Association Award (1950–1951), the Borden Award from the American Home Economics Association (1950–1951), and a research assistantship (1951–1952).

Watts married Charles E. Prothro in 1949. It is said that they met in Connecticut, but this is not clearly documented. Watts Prothro received her PhD from the University of Chicago in 1952. Her dissertation title is "The Relation of the Rates of Inactivation of Peroxidase, Catecholase, and Ascorbase to the Oxidation of Ascorbic Acid in Vegetables."[2] After earning her PhD, Prothro held two positions from 1952 to 1963 at the famed Tuskegee Institute in Alabama: associate professor of chemistry, and professor of home economics and food administration.

Dr. Prothro lived through the racial tension of the civil rights era in Alabama in the 1960s. A train conductor spit upon her for sitting in the "whites only" section when the "colored" section was full. Since she feared for her safety at this time, she left Alabama for the University of Connecticut at Storrs. She became associate professor in the Department of Home Economics from 1963 to 1967. She was there with her husband and daughter. They stayed in Connecticut until her daughter graduated from high school, then moved back to the South.

She returned to Tuskegee University in 1968 and became chair of the Department of Home Economics and Food Administration and research associate at the Carver Foundation until 1980. Prothro was a certified specialist in human nutrition.[3] Her research focused on amino acids and their effect on human nutrition. Prothro's research was in areas of the availability of amino acids from foods and comparison of essential amino acid patterns. She published more than twenty papers.[4]

While teaching at Tuskegee, she mentored many students. She advised a second-year master's student, Bernadine Tolbert, to get a PhD. Tolbert did apply for the PhD program because of this mentoring; she later stated that many

FIGURE 5.1 Johnnie Hines Watts Prothro Credit: Photo courtesy of Spelman College

students respected Dr. Prothro as a "great teacher and trainer: who had 'absolute integrity in collecting and publishing her research.'"[5] Many other students trained by Prothro have gone on to distinguished careers in science.

Prothro received a number of fellowships and grants to study nutrition, among them a National Institutes of Health Fellowship at the University of California at Los Angeles for the study of public health (1958–1959). In 1960, she received three grants from the National Institute of Health, the Human Nutrition Research Division of the U. S. Department of Agriculture, and the National Dairy Council to study human adult male responses to two amino acid patterns. It has been said that "... the impact of her dynamic nitrogen balance research upon the lives of blacks and especially poor blacks living in the deep South deserves recognition among the truly significant contributions of black women scientists."[6]

Prothro also had an appointment as clinical professor in the Department of Allied Health Professions in 1975 and ended her career at Georgia State University as a professor in the Department of Nutrition from 1980 to 1989. When she retired, she "dove into cultural activity, volunteerism and exercise." She awoke at 4:30 every morning except Sunday to walk three miles in the park near her house. She also enjoyed the theater, tended her flower beds, and spent several hours a day at the public library reading for leisure.[7]

Her honors include appointment by President Jimmy Carter to the Board for International Food and Agricultural Development (BIFAD). She was the first African American and the first woman named to this board. Her position on this board involved extensive international travel; she visited Zimbabwe and several West African countries.[8]

She was a member of Sigma Xi, Beta Kappa Chi, New York Academy of Science, Institute of Food Technologists, American Dietetic Association, Sigma Delta Epsilon, and American Chemical Society. In 2003, Dr. Prothro received an honorary degree from Spelman for her research leadership and community activities. Georgia State University sponsors an award for students in her name. Dr. Prothro died of cancer June 6, 2009, at her home in Decatur, Georgia.

RUBYE PRIGMORE TORREY

Rubye Torrey was one of the women featured in the American Association for the Advancement of Science (AAAS) report, "The Double Bind: The Price of Being a Woman and a Minority in Science."[9] She was also one of the first African American women to be active in the American Chemical Society, working on national committees such as the Women Chemists Committee.

Rubye Prigmore was born in Sweetwater, Tennessee, on February 18, 1926, the youngest of three girls. Sweetwater is a small rural town in eastern Tennessee, forty miles southwest of Knoxville. She lived with her mother and grandfather, as her father had died when she was an infant. She credits her grandfather, who was a farmer, for nurturing her love of nature and science. She would follow him around the farm when she was little.

She was educated in the schools of Sweetwater, and her high school chemistry teacher encouraged her to study chemistry. She was very impressed with him. She started college at a small boarding school, Swift Memorial Junior College, a Presbyterian school accredited by the state, for two years.[10] She transferred to Tennessee Agricultural & Industrial State University (now known as Tennessee State University)[11] to complete her college education. The chair of the department of chemistry, Dr. Carl Hill, encouraged her to major in chemistry.[12] Rubye Prigmore graduated with high honors in 1946, and she was fortunate to obtain a scholarship to stay on at Tennessee State to pursue her master's degree, which she obtained in 1948. Her master's thesis project was a project for the Tennessee Valley Authority (TVA).[13] In this project she "developed a chemical method for the quantitative determination of incipient spoilage of fruits and vegetables indigenous to the state of Tennessee."[14] She planned to go on to pursue the doctoral degree in chemistry but decided to work first.

Her first position was assistant professor of chemistry at Tennessee State University, which she held from 1948 to 1963. She then left to pursue her PhD in chemistry at Syracuse University in New York. While at Syracuse, in order to support her degree, she was an instructor in chemistry from 1963 to 1968. She became the first African American woman to receive her PhD in chemistry from Syracuse University in 1968. The title of her thesis is "A Mechanism for the Alpha Radiolysis of Gaseous Hydrogen Sulfide." Her doctoral thesis dealt with the development of a mechanism for the alpha radiolysis of gaseous hydrogen sulfide, which was unknown at that time.[15]

Dr. Torrey (she married in 1957) returned to Tennessee State as a professor of chemistry from 1969 to 1970, when she obtained a postdoctoral collaborative research grant funded by the Atomic Energy Commission to work at Brookhaven National Labs.[16] She was in the mass spectrometry division and looked at the pathways of the gaseous formation of certain noble gases using high-pressure mass spectrometry, a new project at the time. She was there from 1970 to 1974. She was also a visiting scientist at the National Institute of Standards and Technology (formerly the National Bureau of Standards).

At Tennessee State University, she served on the University Research Committee and chaired the Research Committee for the College of Arts and Sciences. In 1978 she founded the university-wide Research Day in order to familiarize students with regional and national research, as students were not allowed to attend scientific meetings and very few faculty members were researchers. In this program, graduate and undergraduate students presented papers before the entire staff, students, and the public. The students were judged on their presentation by a panel of judges from other universities. It has grown since then to become Research Week and now includes invited speakers. Dr. Torrey was honored at the Twenty-fifth Anniversary Research Week event which was held in the spring of 2003. She was also present at the award cermony for the Thirtieth Anniversary Event in 2008 where she gave a talk.

After her position at the National Institute of Standards and Technology, she was hired by Tennessee Technological University as assistant vice president for research and professor of chemistry. Because she did not have a lab at Tennessee Tech, she decided to do research in ethics. She started the Ethics in Science and Technology Division of the Tennessee Academy of Science; this division is concerned with ethics being made an integral part of both research training and the scientific curriculum.[17]

She married Claude A. Torrey, a graduate of the school of Medical Technology of Meharry Medical School, in September 1957. They have two children, Claudia Olivia and Michael Angelo. Currently retired, she has a consulting business. She is a member of the American Society for the Advancemenent of Science, the American Chemical Society, Sigma Xi, and Phi Kappa Phi. She was active in the American Chemical Society as a member of the Women Chemists Committee and the Committee on Chemical Safety.

GLADYS W. ROYAL

Gladys W. Royal and her husband George C. Royal made up a research team who did extensive work on the effect of radiation on human tissue.

Gladys was born Gladys Geraldine Williams in Dallas, Texas, on August 29, 1926. Very little is known about her parents and childhood. She must have entered Dillard University around 1940; she graduated with a BS degree in chemistry from Dillard University in 1944 at the age of eighteen.

In 1947 she married George C. Royal, who was attending the University of Wisconsin, where he received an MS in microbiology. The Royals then moved to Tuskegee, Alabama, where her husband had a position as a bacteriology instructor. Mrs. Royal probably worked in the lab, and in 1954 she received an MS degree in organic chemistry from Tuskegee. She did her predoctoral research at the University of Wisconsin and then applied to Ohio State University, where she was accepted into the doctoral program. The actual date that she received her PhD from Ohio State is not certain but it was either in 1958 or 1959.

A newspaper article about Gladys Royal and her husband states that they were employed by North Carolina A &T (NC A&T). Gladys was a professor of chemistry beginning in 1953, and her husband was a professor of bacteriology in 1952.[18] In order to advance her career, her studies for her advanced degree may have been during the summer or she may have had a leave of absence from her position.

The research project that she and her husband undertook was entitled “Biochemical and Immunological Comparisons of Irradiated Mice and Rats Treated with Bone Marrow Transplants.” The Atomic Energy Commission funded the project, and the Royals were co-directors. The idea of the project was to see if bone marrow transplants could replace tissue damaged by high-dose radiation that was then used in cancer treatment. They wanted to extend this to tissue damaged accidently by exposure to radiation.[19] This research was done during the Cold War, when there was a fear of nuclear war.[20] Their research findings were published in articles at the Fifth International Congress on Nutrition, held in Washington in 1960, and the International Congress of Histochemistry and Cytochemistry, held in Paris the same year.[21] In 1982 Gladys Royal became a biochemist at the U.S. Department of Agriculture Food Service Program, Cooperative State Research Services. She evaluated federal projects in human nutrition.[22]

Royal liked working with students in the classroom; she challenged her students to use their talents to the fullest extent. She urged students to do undergraduate research and in 1970 became president of Beta Kappa Chi Scientific Society.[23] Royal was a member of a number of professional organizations, among them the American Chemical Society, Sigma Xi, and the American Institute of Chemists.

Gladys Royal died on November 9, 2002.

CECILE HOOVER EDWARDS

Cecile Hoover Edwards was a researcher and educator. She devoted her career to improving the nutrition and well-being of disadvantaged people, especially pregnant women. She was cited by the National Council of Negro Women for her outstanding contributions to science and also by the Illinois House of Representatives for "determined devotion to the cause of elimination [of] poverty through the creation of a quality environment."[24] If she were alive in the twenty-first century, she or her students would probably be working on the obesity problem.

Cecile Annette Hoover was born on October 26, 1926, in East St. Louis, Illinois. She was the third child of Annie Jordan, a former schoolteacher, and Ernest Hoover, an insurance manager. She attended the de facto segregated all-black schools of East St. Louis. She graduated with distinction from Lincoln High School at the age of fifteen.[25]

Her choice of where to go to college was made by her mother, an alumna of Tuskegee Institute. She is reputed to have said that one day her mother packed her things and put her on the train for Tuskegee Institute (now Tuskegee University).[26] At Tuskegee, she majored in home economics with a minor in nutrition and chemistry. She later recalled, "I knew from the first day that I had no interest in dietetics. My real interest was in improving nutrition through research."[27] Hoover received a BS degree with high honor from Tuskegee Institute in 1946. She remained at Tuskegee to pursue a master's degree after being awarded a Carver Foundation Fellowship sponsored by the Swift Meat Packing Company. As a graduate student, Hoover conducted the chemical analyses of animal sources of protein. She received a master's degree in chemistry from Tuskegee in 1947. After graduation she received a two-year General Education Board fellowship to study for her PhD degree at Iowa State University. Iowa State had, at the time, one of the nation's first academic departments in foods and nutrition. She majored in nutrition, with minors in physiological chemistry and microscopic anatomy. She completed the requirements for the PhD degree in two years; however, one member of her dissertation committee convinced the others that she was too young to graduate at the age of twenty-two. Instead, she continued her research for an additional year and produced what subsequent students at Iowa State called the "telephone directory"—a dissertation of more than 400 pages, from which she later produced eight publications in peer-reviewed journals.[28] She received a research assistantship from the Iowa State Department of Food and Nutrition for this purpose. The title of her thesis is "Utilization of Nitrogen by the Animal Organism: Influence of Calorie Intake and Methionine Supplementation on the Protein Metabolism of Albino Rats Fed Rations Low in Nitrogen and Containing Varying Proportions of Fat." This dissertation was a study of methionine, an essential amino acid that she said has "not only the good things needed to synthesize protein, but also has sulfur, which can be given to other compounds and be easily released."[29] It is important to note that this research was being done before or at the same time as the discovery of DNA. She would go on to write at least twenty papers about methionine. She received her PhD from Iowa State in 1950.

Upon graduation, she began her career in academia by accepting an appointment as a faculty member and research associate of the Carver Foundation at Tuskegee Institute. She said, "Staying in nutrition at Tuskegee seemed like an opportunity. I felt obligated to pay back the opportunity Tuskegee had given me."[30] In 1952, she became head of Tuskegee's Department of Foods and Nutrition. Her nutritional research later expanded to studies of the amino acid composition of food, the utilization of protein from vegetarian diets, and the planning of well-balanced and nutritious diets, especially for low-income and disadvantaged populations in the United States and developing countries.[31] While at Tuskegee, she obtained a number of research grants. She was the principal investigator on a Carnegie Foundation research project (1951–1952); the director of the Amino Acid Analysis Project; contracted by the Human Nutrition Research Division of the U.S. Agricultural Research Service (1952–1955); the principal investigator on a National Institutes of Health project (1952–1956); and the principal investigator on a Tuskegee Foundation nutrition project (1953–1954).[32] Her research was done either alone or with collaborators at the Carver Research Foundation Laboratories or at the Tuskegee Institute Laboratories. She also taught four courses, and supervised six or more graduate students.

In June 1951, Hoover married Dr. Gerald A. Edwards, a physical chemist who had an appointment at North Carolina Agricultural and Technical State University (NC A&T). Afterward, they worked at Tuskegee Institute and the Carver Foundation until they both took positions at NC

A&T in 1956, where she and her husband began a long research collaboration. They brought with them from Tuskegee a newly awarded National Institutes of Health (NIH) research grant for studies on the metabolism of methionine in the adult rat. That project was supported for the fourteen years that they were at A&T and continued for four additional years after she moved to Howard University. While at NC A&T, Hoover was a professor of nutrition and research. She became chair of the Department of Home Economics in 1968, director of the Undergraduate Research Participation Program in Nutrition (1960–1966), and the director of the Vegetable-Protein Research Project (1964).

In 1966 Hoover took a two-year sabbatical from North Carolina A&T to work in India. She brought one of her research projects with her and worked at the Central Food Technological Research Institute in Mysore and conducted other studies relevant to nutrition in India.

In 1971, Hoover joined the faculty at Howard University as a professor of nutrition and chair of the department. Her first task at Howard University was to design the structure of the new school of Human Ecology, working with the faculty to identify its curricula and programs. She was the dean of this school from 1974 to 1986. This program attracted a committed faculty, 82 percent of whom held the PhD degree. She also established a PhD program in nutrition at Howard, which was the only such program at a predominantly black university in the United States.

She was also interested in conducting research aimed at improving the quality of life for minorities and the socially and economically disadvantaged in this country and developing countries. She coauthored a book with other members of her faculty entitled *Human Ecology: Interaction of Man with His Environments.*[33] She assembled a team of researchers at Howard to engage in research on African American women and their children entitled, "Nutrition, Other Factors, and the Outcome of Pregnancy." The proposal for this research received an award of $4.6 million from National Institute of Health (NIH). It also generated a special supplement to the June 1994 issue of *The Journal of Nutrition*, which was devoted entirely to the first thirteen papers reporting the results of this research. Dr. Edwards was guest editor of this special supplement.[34]

During her tenure at Howard, Edwards also served as dean of the School of Continuing Education (1986–1987), and interim dean of the new College of Pharmacy, Nursing, and Allied Health Sciences (1997–1998). Edwards retired from Howard in 1999 as Professor Emeritus of Nutrition.

Over the course of her long career, Edwards published over 160 articles in peer-reviewed journals and received numerous honors from Iowa State University, Tuskegee Institute, NC A&T and Howard University. She was an excellent teacher, stressing both the curriculum and the individual value of each student. She emphasized the need for all students to achieve excellence in their professional endeavors. She was a good role model for all her students and faculty.

At home, she was an excellent cook, an engaging hostess, and wonderful wife and mother. She and her husband had three children, Jerry, Adrienne, and Hazel. Each of them graduated from Howard University and went on to graduate degrees in the fields of business, veterinary medicine, and architecture. They also had a grandson, John, who graduated from Howard. Her husband, Gerald A. Edward, died in June 2005, not long before her death on September 17, 2005.

ALLENE JOHNSON

Allene Johnson is a chemist who chose to pursue the field of chemistry by teaching high school chemistry. She was born on April 20, 1933, in Supply, North Carolina, now called Shallotte. She was one of six children. Her mother and father were both farmers. The first school she attended was Cedar Grove School in Supply. The schools she attended were segregated, since she grew up before the *Brown v. Board of Education* decision. There were no specific science courses taught in her early school years, and her formal study of science began in the sixth or seventh grade. There were no laboratory experiments, but she learned science creatively while working on the farm and building her own "toys." She attended Brunswick County High School in Southport, North Carolina. It was there that she began her study of science, as it was a required subject for all four years. She graduated in 1950.

Her parents believed in education for all of their children. It was understood that she would attend college. She attended North Carolina College (now North Carolina Central University), a historically black college.[35] She was not able to attend a college out of state because her family could only afford in-state tuition. Johnson chose her major in chemistry while attending a career fair at the college. She had to decide between math and science; she chose chemistry, with mathematics as her minor. She started her major in her freshman year, since that was the norm for students when she went to college; she was one of only two women majoring in chemistry that year. Allene graduated from North Carolina College with a BS degree in 1954.

Johnson's career goal was to go into industry after graduation. She needed to work instead of going directly to graduate school in order to support her other siblings in college. She received a job offer with the Tennessee Valley Authority (TVA). (This was 1954, when the federal government was the major employer of African American chemists who did not pursue an academic career.) She did not take the TVA job because she wanted to stay in North Carolina; therefore, she chose teaching, beginning her career in 1954 by teaching chemistry, physics, and math at Union High School in Shallotte, North Carolina. This was the same year as the *Brown v. Board of Education* ruling, but the schools did not become integrated right away. She had no equipment or chemicals in her school but she was able to improvise. She bought her chemicals and equipment from the drug store and used what she could find in her environment.

After the Soviet launch of Sputnik,[36] the National Science Foundation set up graduate courses for teachers, and Johnson took graduate courses for three years at North Carolina Central University. She applied and received an academic year appointment to Ohio State University, where she received a master's degree in chemistry. She continued to teach at Union High School in NC until 1963.

Her brother worked as a chemist in New Jersey and convinced her to move there so that the family could be close together. She passed the New Jersey examination for a teaching certification and applied to teach in Summit High School in 1963.[37] She became the first

African American schoolteacher in the Summit New Jersey School District, and she was hired to teach chemistry. She received the job not because of affirmative action, but because she had the best credentials as a chemistry major with both a BS and an MS in chemistry. She continued to hone her skills as a chemistry teacher by participating in numerous summer workshops for teachers.

Even though she was teaching in a school system where the majority of students were white, she did not abandon her commitment to minority students. She became the magnet for all of the minority students in the high school, even if they were not taking chemistry. They came to her for advice and mentoring. Many of her students became doctors, chemical engineers, and microbiologists. Johnson retired from teaching in 1997 but continues to do volunteer work for science education.

Johnson believes that her greatest contributions to the minority community came through her volunteer activities. She was a tutor and participated in the Summer Science School at Bell Laboratories in Murray Hill, New Jersey. She now works with minority youth through the sorority Delta Sigma Theta. The national organization of the sorority received a grant from the National Science Foundation to found a Delta Academy for math and science.[38] Johnson works with the North Jersey Chapter of the Deltas in the science area; the program is for minority middle-school girls. She also conducts "Chemistry for Kids" workshops for students at local libraries, sponsored by local minority groups. These workshops are designed to enhance the education of minority students.

She is an active member of the North Jersey section of the American Chemical Society's Project SEED committee.[39]

She is the person who is responsible for going to the schools in Newark, Irvington, East Orange, and Orange, New Jersey (predominantly African American school districts); she has been able to increase the number of African American students who participate in Project SEED, and has mentored other minority students as well.

Johnson also diligently works to increase the number of minority chemistry teachers and middle-school science teachers. Together with Dr. George Gross, another award-winning educator member of the North Jersey American Chemical Society, she has done numerous teacher workshops for the teachers of the Newark School district (one of New Jersey's largest minority districts, with both African American and Hispanic students). Teachers who give back to other teachers have the capacity to increase the number of minority students in science.

Johnson is the recipient of many local, state, and national awards, including the Presidential Award for Excellence in Science and Mathematics Teaching, American Chemical Society Middle Atlantic Regional Award in High School Chemistry Teaching, National Catalyst Award from the Manufacturing Chemists Association, and, most recently, the New Jersey Science Teachers Association (NJSTA) Citation Scroll, which is the NJSTA's highest award. Here is an excerpt from the NJSTA award address:

> It is Allene's human qualities, however, that people are most apt to speak about. She is known for being a champion of the disadvantaged, regardless of their background. She gives a voice and support to people who need an advocate. She is respected for her work with Project SEED, an American Chemical Society (ACS) initiative designed to encourage economically disadvantaged high school students

to pursue career opportunities in the chemical field. She also received the 2006 ACS Stanley Israel Award for advancing Diversity in the Chemical Sciences.

Ms. Johnson never married, choosing to devote her life to her career, family, and the mentoring of students and teachers.[40] She was a member of the American Chemical Society ACS Secondary Exams Committee and has served as both secretary and chair for this committee. She also served on the ACS Exams Institute Board. She is coauthor of a laboratory manual and worked on a book, "Laboratory Assessment Builds Success," by the Institute for Chemical Education.

MARY ANTOINETTE SCHIESLER

Dr. Antoinette Schiesler started her career as a Oblate order nun and elementary school teacher; after leaving the order, she became a college administrator and champion of improvement of science education for minority students.

Schiesler was born Carole Virginia Rodez in Chicago, Illinois, on December 13, 1934. Her mother Gladyce was a singer and had traveled to Chicago from her hometown of New Haven, Connecticut, in order to find work with the big bands. It was there she met a Cuban man named Rodriquez, and Carole was the result of this first encounter. Carole did not know this until later in her life; she lived with her mother as a single parent in New Haven, where her mother had returned because she had more friends and relatives who could help take care of Carole while she worked. They lived an unsettled life as her mother struggled to get jobs to live and pay the rent. Carole was sometimes placed with friends and relatives while her mother worked out of state. She was a bright child, interested in science and how things worked. She took one of her dolls apart to see how it made noise. One of her favorite birthday presents was a chemistry set. The first thing she made with the set was purple ink. Her mother encouraged her by taking her to free concerts in the park, to libraries, and on tours on the public bus, where she learned to read by reading the advertisements on the bus.

When Carole was twelve years old, her mother decided that she should have a stable life and a good education. Gladyce valued education, having had two years of college at City College in New York. Therefore, Carole was placed at St. Frances Academy, a Catholic boarding school for "colored girls." She was not Catholic at the time, but that did not matter. The Oblate Sisters of Providence, free black Roman Catholic nuns, primarily of Haitian descent, had founded St. Frances Academy in Baltimore in 1828.[41] At that time, black children were not allowed to be taught in Baltimore. St. Frances became the first school for black children in the city. It was a school for girls in grades eight to twelve.[42]

Carole fell in love with the school, its discipline, and its nuns. At first, she resisted because she was not Catholic, but later decided she wanted to become a nun and converted to Catholism after the tenth grade. During her last years at the school, she became a scholarship student as her mother lost her job and could not afford to pay the tuition. She was different from most of the girls at the school because of her receipt of a scholarship.[43]

She was required to take chemistry as a senior at St. Frances, and she loved it. She became the teacher's

FIGURE 5.2 Mary Antoinette Schiesler Credit: Photo courtesy of the Oblate Sisters of Providence Archives, Baltimore, MD

favorite and was given challenging assignments. She also started to run the lab and teach other students when the teacher was absent. She became the best in the class, and she loved the structure and discipline of chemistry, paired with creativity and discovery.[44] The discipline also appealed to her stage of life. She explains, "I was looking for answers as an adolescent, and chemistry provided them. I was extremely scrupulous."[45]

She entered the Convent of the Immaculate Conception on September 8, 1952, at the age of seventeen, as Carole Rodez. She became a postulant with the new name of Mary Antoinette on March 9, 1953, and would retain this name for the rest of her life. She took her final vows on March 9, 1955.

After two years of teacher training at the Oblate Institute, she was sent off to teach elementary school in Washington, D.C.; she hated it because she had not had classroom management training and did not know how to handle the students. Her third assignment was teaching science and math to the older kids, and she loved it. She was going to the College of Notre Dame of Maryland part-time while teaching, and full-time in the summer. She wanted to go to school full–time, but that was not possible. She received her BA degree in 1967.

Because of her skills in science, it was suggested that she apply for a program for junior-college teachers of math and science sponsored by the National Labs in Oak Ridge, Tennessee. Mary Antoinette was accepted, even though she was an elementary school teacher. Only twenty students had been accepted into the program, and she was the only African American student. She received grudging permission from the Mother Superior to go, even though she should have asked for permission before she applied to the program.

When she arrived at Oak Ridge, it was the first time that she was on her own in an apartment. Living in a convent would have been too much because of her exhausting schedule. She taught herself some of the prerequisites to the curriculum, because she had never had physics and she was to learn nuclear physics. It was at Oak Ridge that she became self-confident as a student and thinker. She enjoyed the making of science.

Her advisor at Oak Ridge was a biochemist. Even though she had not studied biochemistry, he became her advisor by default. Her advisor made her work on a piece of his research as a thesis topic and told her that she was on her own in doing so. Her topic became the effect of radioactivity on enzymes. She had to learn everything from the beginning. She succeeded and wrote a master's thesis entitled "The Inactivation of Pancreatic Lipase by Gamma Radiation."[46] She received her master's degree in August 1969 from the University of Tennessee, Knoxville, having achieved it mostly by independent study.

After this experience at Oak Ridge and the taste of independence, Mary Antoinette went back to the convent, but life was not the same. She returned to Mt. Providence Junior College as registrar, academic dean, and instructor in math. She had been a person who had loved being a nun, but not anymore; after a year, in 1971, she left the order. This was the only reason that the Mother Superior had not wanted her to go to Oak Ridge: she felt that Mary Antoinette might leave the order. Mary Antoinette had been a good nun, so good that she might have been in line to become the next Mother Superior, but she gave up that calling.[47]

Soon after she left the order, she met Robert Alan Schiesler, a former Roman Catholic seminarian. He had left the seminary, taught school for a while, and decided to become an Episcopal priest. They were married on October 20, 1973, even though he was a white man who was fourteen years younger than she was. The couple ran into trouble when they traveled in the South, as in some states it was still illegal for a white person to be married to a black person.[48] They remained married until her death in 1996. She resisted being a clergy spouse (who served tea and cookies to the women in the church, etc.), as she was a professional and wanted to remain so.

Her first job on her own was co-teaching a course at Bowie State College, a historically black college. She then went to the University of Maryland, College Park, to pursue a PhD in chemical education. She worked at the University of Maryland in the school of Pharmacy as a lecturer and lab coordinator. She set about to rewrite the workbooks in math and science because "they were not good." She had a series of jobs at the Maryland State Board of Higher Education; the National Science Foundation, program manager; Eastern Michigan University, director of research; Villanova University, director of research and sponsored programs; and finally, in 1988, academic dean at Cabrini College in Radnor, Pennsylvania, until 1993.

She decided to become an Episcopal priest. Her husband, also a priest, did not want her to do so at first; he felt she was too old at fifty-two to begin a new career and that she would take a major cut in salary. He apparently was also concerned about competition between them. She decided that her ministry would be different. She had started a women's spirituality group at Cabrini College, in which women met once a week to "discover the spirit." She had long ago learned to play a guitar and sing, which she did at these events. She did not want a hierarchical group, but more of a circle of equals. She was ordained and became the assistant to the dean of the Cathedral of St. John in Wilmington, Delaware, until her sudden death, caused by a brain tumor, in 1996. Her family has sponsored a scholarship at Cabrini College in her memory.

Throughout Dr. Schiesler's life, she was able to join two disciplines, science and religion. Even now, scholars are trying to correlate religion and science.[49] To be a scientist you must adhere to the principals and "rules" of science. Science is based on observable, testable data; some people say that religion is based on beliefs that have no concrete or tangible evidence. She did not see it that way. She believed that religion and science reinforced and strengthened one another. She stated, "The beauty of chemistry and astronomy say so much to me about the beauty and wonder of God. . . . How could you take a chemistry course and not believe in God? . . . the order, the fitting together . . . all God's work. Chemistry is the microworld, disciplined and ordered, and astronomy is the macroworld, harmonious and ordered."[50]

GLORIA LONG ANDERSON

Gloria Long Anderson is a pioneer among women chemists in the field of fluorine 19 organic synthetic chemistry and nuclear magnetic resonance (NMR).[51]

Gloria Long was born on November 5, 1938, the fourth child and only girl in a family of six, in racially segregated Altheimer, Arkansas. Her father, Charley Long, was a sharecropper, and her mother, Elsie Lee Foggie, was a homemaker. Her father only had a third-grade education, while her mother had completed the tenth grade. She was raised in a farming family in which everyone was expected to work. . Long later recalled, "In those days we didn't know we were living in poverty."[52]

A lover of books, Long learned to read by the age of four and entered school at that age. She later attended her town's all-black high school, Altheimer Training School. The school's faculty was all black, and they pushed all their students to excel and succeed, despite the many obstacles they knew their students would face. Despite their sharecropper status, Long's parents wanted their children to succeed academically.

Long graduated from Altheimer Training School in 1954 at the age of sixteen. Refusing to remain in Altheimer and work as a domestic for white families (one of the few employment options for black women in the town), Long chose to enroll in college and study for a career. She thought that she might enroll in a pre-med program. The only college that she could go to in Arkansas at that time was Arkansas A&M Normal College, the state-supported school for African Americans. It was close to where she lived and almost affordable. Because she had good grades, she received a small scholarship the first year. Starting with her second year of college, she received a Rockefeller Scholarship that funded the rest of her college career. She decided to major in chemistry and received excellent grades. When she graduated in 1958 she graduated *summa cum laude*, first in a class of 237 with a GPA of 2.96, on a 3.00 scale.

After graduation from college, she was expected to return home and teach, since the only two options of employment at that time for educated blacks were preaching and teaching. In fact, when she graduated from college, she was the only one in her class without employment plans. She applied to Stanford University for graduate school and was admitted but could not attend without funding. She applied for a position at Ralston Purina as a chemist at that time but they did not hire her. (Industry was slow to open positions to chemists at this time before affirmative action.)[53] Forced to teach, she taught seventh grade for six months.

Gloria Long benefited from the new emphasis on math and science in the United States after the launch of Sputnik in 1957 by the Soviet Union. Atlanta University had one of the new National Science Foundation grants for black high school math and science teachers. Long was contacted by Kimuel Alonzo Huggins, chair of the chemistry department of Atlanta University, to sign up for this program for a master's degree in chemistry. He also arranged for the funding. Gloria entered the program in 1959 with the intention of going on to receive her MS in chemistry. However, in 1960 she married Leonard Sinclair Anderson, another schoolteacher, and she found that it would have been difficult to continue financially in spite of the fellowship. That is when Dr. Huggins stepped in as a mentor, found research money for her and begged her to continue her studies. It was while studying for her MS degree that she fell in love with organic chemistry. She completed the requirement for an MS degree in organic chemistry in 1961. Her master's thesis title, "Studies on 1-(4-Methylphenyl)-1,3-Butadiene," described a new three-step synthesis of butadiene, a gaseous hydrocarbon. The structures were confirmed using infrared spectrometry. In 1961, she also gave birth to her son Gerald.

Armed with her MS degree in chemistry, she accepted a position to teach at South Carolina State College in Orangeburg, South Carolina. She stayed there for a year, until Dr. Henry McBay lured her back to Atlanta at Morehouse College. Dr. McBay has been called the "father of black chemists in the United States" because he trained the lion's share of black chemists between the years 1936 and 1994.[54] He was aware of Gloria's thesis work at Atlanta University and he arranged for her to come to Morehouse as a chemistry instructor and research assistant. She spent two years at Morehouse and was encouraged to pursue the doctorate degree if she was to continue teaching at the college level.

In the fall of 1965, she enrolled at Dr. McBay's alma mater, the University of Chicago. She received a research and teaching assistantship. The University of Chicago has an excellent record of producing black science doctorates; however, Anderson felt that there was racism among some of the faculty.[55] Even though there was no formal mentor or support program, she did find a role model, Thomas Cole, a teaching assistant in the instrumental course. Impressed with his ability with instruments, she felt that it offered proof that a black person could do that type of work.[56] During her first year in graduate school, she started tutoring white female undergraduates, but stopped when she realized they had advantages she did not have. She had two women study partners, one of whom, although white, became her friend. She later dropped out of the program, leaving Gloria Long Anderson to study on her own.[57]

Anderson's dissertation project was a study of fluorine-19 nuclear magnetic resonance (NMR) substrate chemical shifts and CF infrared frequency shifts, and her PhD professor was Dr. Leon M. Stock. Her research generated at least one paper, which was published before she had completed her dissertation. (The "19" following fluorine refers to the isotope of [the] element fluorine that, like other elements with odd numbered masses, has magnetic properties.) Fluorine-19 chemistry became an important field of research shortly before World War II, when many commercial uses for fluorine compounds were discovered. Much of Anderson's research has involved nuclear magnetic resonance (NMR) spectroscopy, a method of investigating organic compounds by analyzing the nucleic responses of molecules subjected to radio-frequency radiation with a slowly changing magnetic field. NMR spectroscopy, which has been widely exploited for chemistry. NMR spectroscopy enables extremely sophisticated analysis of the molecular structures and interactions of various materials. Fluorine NMR spectroscopy is now being applied to a range of biochemical problems, including the study of the human metabolism and the formulation of new pharmaceuticals.[58] At the time she was doing this work, she was one of the pioneers in the field and the only one in her university research group who knew how to switch the NMR Instrument probe to F-19.

Shortly before Anderson received her PhD, she met with her mentor, Dr. McBay, who told her to apply for the position as chair of the Department of Chemistry at Morris Brown College in Atlanta. Dr. McBay is alleged to have said, "Morris Brown College needed a chairman of chemistry department . . . there is no reason that you should not be the chairman of that department."[59] This is unheard of, as young, newly minted PhDs are never appointed to the position of chair of a chemistry department right out of graduate school. She graduated from University of Chicago in 1968 and became an associate professor and department chair at Morris Brown. This is an example of the power and influence that Dr. McBay had.

Dr. Anderson really wanted the opportunity to work at a small black college. While she was still at the University of Chicago, Dr. Martin Luther King was assassinated. She pledged that her contribution to a continuing civil rights movement would be to teach only in a small black college. As a researcher, she also knew that would possibly be denying herself access to funds that she needed to continue her research. In addition to her work at Morris Brown, Anderson found time to do post-doctoral research at Georgia Institute of Technology under Dr. Charles L. Liotta. The title of the work was "Studies on the Mechanism of Epoxidation."

Anderson's research involved not only the synthesis of complex organic molecules but also the structural analysis of the compound and how they were made. She said, "If a hydrogen group is replaced with another chemical group, for example, how does it transmit its effect to the reaction site? It has to transmit its effect to the site, the real interesting question is how?"[60]

She continued the research about F-19 organic compounds that she had begun in graduate school. She was, therefore, able to understand how chemical substituent's combined at the molecular level and to develop synthetic techniques for developing solid-fuel rocket propellants, synthesis of antiviral drugs and medicinal compounds containing fluorine, structural studies on substituted amantadines, and studies on the mechanism of epoxidation.[61]

In addition to her own research, her position as chair of the Department of Chemistry required her to apply for research grants. She applied for and was granted a mainstream National Science Foundation grant, but it was not funded for a year. She found out later that her grant was funded from the funds allocated to minority institutions, instead of funding her mainstream grant. When she applied for renewal of the grant, they cut her off without supplying her phase-out funds, as they usually do to mainstream institutions. She attributes that to the racism of the time. When she applied for funds under the minority grant process, she was told that she was not eligible because she had received a mainstream grant.

Undaunted, Anderson applied for funding from other sources and received funding from Lockheed Corporation. However, this meant that she still could not get an NSF grant. She thinks that had she been at a major white university, she would have received funding for equipment and graduate students to help, and would have had faculty colleagues. She currently does her research on her own, with an occasional undergraduate research student. She has been able to secure more than a quarter million dollars in research grants through 1985. She also funds her research with money from her own salary.

Anderson has numerous publications and numerous patents currently pending. She is a consultant for Bio SPECS, a chemical and pharmaceutical research firm in the Netherlands, which has purchased a number of her compounds to use as potential targets for research. She either holds the patent on these compounds or has applied for the patent so they cannot be worked on without her approval. She also had a number of compounds prepared for their antiviral properties but thus far these have not been tested.[62]

She continued to have a heavy teaching load in addition to research because of her commitment to students. In her position as chair of the Chemistry Department at Morris Brown, she improved the chemistry department by hiring five chemistry professors with PhDs and by upgrading the program so that it would be approved by the American Chemical Society, with the introduction of modern scientific instrumentation to the chemistry labs.

In 1973, she became the Fuller E. Callaway Professor of Chemistry.[63] She held that position until 1984, when she left the Department of Chemistry to become dean of academic affairs. She resumed the post of Callaway Professor in 1990 and currently holds that position.

Her awards include, among others, Outstanding Teacher at Morris Brown (1976), Scroll of Honor from the National Association of Negro Business and Professional Women (1983), Teacher of the Year Award (1983), and inclusion in the Faculty/Staff Hall of Fame.

In 1972, President Nixon appointed her to a six-year term on the board of the Corporation for Public Broadcasting (CPB). She chaired the committees on Minority Training, Minorities and Women, and Human Resources Development. She was vice chair of the CPB board in 1977–1979. Her job there was to ensure the diversity of programming in public broadcasting. She is a member of the American Chemical Society, the National Institute of Science, and the National Science Teachers Association, among others.

Dr. Anderson's current position (2011) is assistant to the president of Morris Brown College, which has been struggling since losing its accreditation due to mismanagement of funds; most of its staff has since left. Dr. Anderson still teaches a non-laboratory course in order to remain Callaway Professor.

When speaking to a group of young scientists, she said, "[You] can do anything that you want to do. You can be anything that you want to be. However, you must be determined. You must work hard. You must not let anyone define who you are or what you can do." She went on to say, "As Dr. Martin Luther King, Jr. said at my college commencement, and I paraphrase, 'Don't go out to be the best black scientist, Go out to be the best scientist.' "[64]

LINDA C. MEADE-TOLLIN

Dr. Meade-Tollin is a biochemist whose research interests have included angiogenesis, matrix metalloproteinase expression, and cancer-related invasion and metastasis. She has developed and modified an in vitro assay for angiogenesis in human microvascular endothelial cells, which was used to identify and purify potential angiogenic inhibitors and enhancers from desert fungi.

Linda Meade was born in her parent's home in London, West Virginia, on August 16, 1944. She was the second child in the family, which included an older brother, Robert Alfred Meade III. Her father, Robert Alfred Meade II, was the area's only African American dentist; he had obtained his DDS degree from Howard University in 1926. Her mother, Virginia May Daniels, was a graduate of West Virginia State College who taught English as well as French and Spanish in West Virginia secondary schools. She also earned a master's degree from Columbia University in counseling. The family had settled in London, West Virginia, a small rural town of approximately three hundred people in the 1930s. The area was only 3 percent black; many of the people were coal miners.

In spite of the small black population in the county, the schools were segregated until Linda entered the ninth grade. She attended Washington Elementary School, a two-room school with grades one through three in one room and four through six in the other. When she entered school, since she had essentially been homeschooled and already knew how to read and write, her teacher refused to teach her and insisted that she be assessed. She was evaluated by the county educational assessment division and tested on a junior high school level. She was placed in the third grade for a year to mature with her age group and then promoted to the fourth grade the following year. She had only black teachers in that school and very little science in the fourth through sixth grades. She then attended segregated Washington Junior High School, which shared the building with the elementary school, where she followed a college-track curriculum that included biology.

When she completed the eighth grade, West Virginia schools were required to be integrated. This resulted in her transfer to Cedar Grove High School in 1956. She was bussed six miles to Cedar Grove High School[65] in Cedar Grove, West Virginia, for the remainder of her high school attendance. There she continued to follow a college preparatory curriculum that included biology, chemistry, physics, and French. The black students who were the first to attend the formerly all-white Cedar Grove High School excelled in their studies because they had been well taught by their black teachers.[66] The first year after integration there was a black valedictorian and a black salutatorian. The year Linda graduated, the top three graduates of Cedar Grove High School were black women. One of her role models and inspirations at Cedar Grove High School was Henry Davis, a master teacher, slight in stature, with an amazingly positive attitude; he seemed to have an unlimited store of mnemonic tricks for every subject. He taught chemistry, physics, and French.

Linda graduated as salutatorian of her high school class in 1960 at the age of sixteen and applied to Oberlin College, Radcliffe College, and several others. However, because she was so young, her parents sent her to her mother's alma mater, West Virginia State College (now West Virginia

State University), a historically black college that was about thirty miles away. Her mother had many friends and colleagues there who could keep an eye on Linda and provide guidance and support. Even though West Virginia State was a historically black college, it called itself "a living laboratory of human relations" because the integration process took place smoothly. The majority of the students were white and commuted to the school, while the black students lived on campus in the dorms; so, the college appeared predominantly white by day and predominantly black by night. At West Virginia State, she majored in chemistry and minored in biology and mathematics. She loved science, so she decided to major in chemistry, but she since also liked other areas just as well, so she minored in biology and mathematics. Chemistry was the most difficult subject but she put in the effort required to be successful. At graduation, there were four surviving chemistry majors of the 144 declared majors in her freshman class, two black women and two white men who studied together and often prayed together for success on the organic chemistry exams. She also had time to enjoy the social life of the campus by joining Beta Kappa Chi, the college marching band, and AKA (Alpha Kappa Alpha), a black women's sorority.[67] In 1964, when she was nineteen years old, she graduated cum laude from West Virginia State with a BS in chemistry.

She decided to go to graduate school because she enjoyed learning and felt that the life of a scholar would never be boring. She was fascinated with how and why biological processes functioned. Then, she had to decide what to study in graduate school. Since there were not many opportunities for undergraduate research in the sciences at West Virginia State, she felt that research would be a good area for lifetime learning. She was drawn to different aspects of both chemistry and biology, and so decided to combine them and study the developing area of biochemistry. She had never had a course in biochemistry but she knew it was the marriage of biology and chemistry and had to do with how living cells functioned. Since she was still young, she decided to go to graduate school in New York because she could initially live with one of her mother's sisters, Alberta Campbell. She worked for a year at Harlem Hospital in order to get a break from college. It also gave the country girl time to transition to living in a big city.

She applied to only one graduate school, the more affordable City University of New York (CUNY). She was accepted in 1965 to the doctoral chemistry program at City College (CCNY). Although, there was no doctoral biochemistry program at the time, she was advised that she could focus on a biochemical area within the chemistry program and achieve the desired training. Soon afterward, a doctoral biochemistry program was established at CUNY, and the new program was more suited to her interests. She transferred to that program and became one of its first graduates. The City University of New York had several separate campuses located in each of the boroughs of New York City, in addition to the central graduate office in midtown Manhattan, making it possible to attend classes and work in laboratories at different sites. The first year she attended classes and held a teaching assistantship at City College. The second year she was awarded a research assistantship at Hunter College with Dr. Maria Tomasz. Her thesis advisor was Dr. Burton Tropp, a new faculty member at Richmond College (now merged with Staten Island Community College to form the College of Staten Island).

FIGURE 5.3 Linda C. Meade Tollin and G. Tim Bowden Credit: Photo courtesy of G. Tim Bowden

When Dr. Tropp transferred to the Queens College campus, she traveled with him as his first graduate student.[68]

She lived in the heart of Harlem while in graduate school to draw strength from proximity to her culture, but she took classes and did laboratory work at Hunter, Richmond, and Queens Colleges, and at the Graduate Center for the City University on 42nd Street in Manhattan. Laboratory and library research often went on all night until early morning, so living in Harlem and going to school at the different campuses was a physical as well as mental challenge. Traveling late at night alone on subways, buses, ferries, and city streets was potentially dangerous. She would often leave the subway stop near her house at two or three in the morning with heavy books and run to her nearby apartment complex, past the doorman, and not stop until she got on the elevator. This was not her only problem while in college. It was 1965, the beginning of affirmative action at City College. Resentment to the admission of supposedly unqualified minority students to the highly competitive environment at CCNY was present. Seeking a supportive environment, she explored transferring to a prestigious doctoral program at another institution. There an interviewer asked her if she ever planned to get married, because the program did not want to invest time in training a woman who would get married and not use her skills. This was an example of the double bind of being a woman and black in science.[69] In the process of exploring other academic options, she met and was advised by Dr. Marie Daly, a well-respected black female biochemist who was an inspiring role model; it was Dr. Daly who referred her to the interviewer.[70] Dr. Daly was encouraging and supportive of her and was the stepmother of one of Linda's best friends at the time, Rebecca C. Smith.

After weighing different options, she decided to remain enrolled in the City University. She realized that there were difficulties to be overcome no matter where she was. At CUNY, she was able to take courses taught by City University and Columbia University faculty and researchers, including lectures by Nobel laureates. She earned an MA degree in biochemistry from Hunter College in 1969 and a PhD in biochemistry in 1972.[71] The title of her thesis is "Metabolic Consequences of the Relaxed Gene in E. coli." She investigated cellular effects of the altered expression of a bacterial gene.

After graduate school, Meade really wanted to pursue an academic career involving both teaching and research in a historically black college because she wanted to give back to her community. Before she received her PhD degree, she taught introductory chemistry courses to many minority students in affirmative action programs both at CUNY campuses and at Barnard College. She accepted a tenure track assistant professorship at the college of Old Westbury, a branch of the State University of New York. Since research facilities were not available at Old Westbury, as promised, and despite a fifteen-contact-hour teaching load, she accepted a visiting assistant professorship at Rockefeller University in the laboratory of Dr. Anthony Cerami, a sickle cell anemia researcher. She loved working at Rockefeller University because of both the cutting-edge research performed there and the exposure to a combination of basic research with clinical practice. She managed both research at Rockefeller and teaching at Old Westbury by arranging her schedule so that her teaching responsibilities fell on three days each week at old Westbury and then spent the remaining days in laboratory research at Rockefeller.

She then decided that she preferred to do pure research and teaching at the same institution. To enhance her research training, she applied for and in 1975 was awarded a National Institutes of Health postdoctoral fellowship at the University of Arizona. While at Rockefeller, she was introduced to another visiting professor, Dr. Gordon Tollin, by a colleague, Dr. George Holmes. Gordon Tollin was on sabbatical leave from the University of Arizona and was a visiting professor at Rockefeller University. The two convinced her that she would find a productive and supportive research environment at the University of Arizona. She successfully competed for an NIH National Research

Service postdoctoral award and joined the laboratory of Dr. Christopher Mathews in the Biochemistry Department at the University of Arizona College of Medicine. Dr. Meade and Dr. Tollin were married in 1978. Dr. Gordon Tollin is a highly respected, internationally known biochemist.

Dr. Meade-Tollin remained at the University of Arizona until 1986. She also became the primary caregiver for her parents, who came to live with her shortly after she was married. These personal commitments made it difficult to be mobile enough to enhance her career opportunities, and in some sense, she was a captive spouse. In 1984 her daughter Amina was born; because of this additional responsibility at home, she decided she would not actively pursue a tenured position at the University of Arizona. Dr. Meade-Tollin filed a suit against the University of Arizona, alleging discrimination in hiring; during the negotiations, she accepted a faculty fellowship (1985–1986) at Morehouse Medical School in Atlanta, Georgia. At Morehouse, the first all-black medical school, she focused on the students. Students were supported and nurtured at their particular level of ability and were brought to academic excellence. She taught and did research at Morehouse for the academic year, but then returned to the University of Arizona and an acceptable legal settlement.

Although she had chosen not to pursue a tenure track position at the University of Arizona, she held a number of teaching, administrative, and research positions. For many years, she was the only African American woman to head an independent research lab at the University. She received a five-year career development award (1996–2001) from the National Institutes of Health, and received the Henry H. Hill Lecturer award for outstanding contributions to the field of biochemistry and medical research from the National Organization for the Professional Advancement of Black Chemists and Chemical Engineers in 1998. She served for many years on the editorial board of the *Acta Histochemica*, an international scientific journal, has authored fifteen peer-reviewed journal articles with nationally and internationally recognized colleagues, and coauthored an introductory chemistry curriculum based on nutrition.

Although she loved teaching, she preferred research. The research project of which she is most proud is her development of a more physiologically relevant and highly reproducible rapid model angiogenesis assay utilizing human microvascular endothelial cells. This system was then used as a drug discovery approach to screen over a thousand extracts of desert plants in a successful search for compounds with anti-angiogenic or pro-angiogenic activity. Angiogenesis is defined as the growth of blood vessels. These blood vessels can supply nutrients to the tumor, remove waste products, and serve as a highway for cancer cells to metastasize (i.e., spread to distant sites throughout the body). Angiogenesis is thus a critical aspect of cancer invasion and metastasis and a significant therapeutic target. This work was done while she was a member of the Arizona Cancer Center and the Department of Surgery at the University of Arizona College of Medicine, in collaboration with Dr. Leslie Gunatilaka, a chemist, and Dr. Luke Whitesell. A similar assay has been independently commercialized and is sold as a kit by Becton Dickinson.

Meade-Tollin is also proud of her ability to work with majority and minority students, training them in research techniques and concepts. She has worked with many students who had no previous laboratory research exposure in keeping with the values practiced by minority institutions. Many of these students continued on to careers in biomedical research and other professional and graduate degrees. She was also involved in mentoring women and encouraging them to pursue careers in science. She was the director of the Office of Women in Science and Engineering that is part of the Department of Women's Studies at the University of Arizona. She organized conferences and activities to encourage women to go into science and engineering, and has assisted in presenting leadership workshops for women in medical academia for the American Association of Medical Colleges.

Among her volunteer activities, she was active for many years in the National Organization for the Professional Advancement of Black Chemists and Chemical Engineers (NOBCChE). She was the first woman to chair the Executive Board of that organization. She was delighted to work with NOBCChE because it provided an opportunity to interact with other black scientists and science students nationwide.

Dr. Meade-Tollin is currently divorced but resides in Tucson, as do her brother, daughter, grandchildren, and ex-spouse. Dr. Meade-Tollin retired in the spring of 2008 as a Research Assistant Professor Emerita. She is currently the director of Anti-Metastasis Research for CCW Corporation, a newly established biotech company whose goal is to cure cancer worldwide, and also is pursuing her interests in travel, spirituality, family interactions, and Wu style Tai Chi.[72]

LYNDA MARIE JORDAN

Lynda Jordan, biological chemist,[73] a participant in the PBS documentary series *Discovering Women* by WGBH in Boston, has made outstanding accomplishments in science.[74] This pioneering award-winning series profiles the lives of six women scientists. The one-hour segment that profiles Jordan's life and the significant accomplishments of her early career is titled "Jewels in a Test Tube."

Lynda was born in Boston City Hospital, located in Roxbury, Massachusetts, on September 20, 1956. She is the oldest of three girls born to Charlene Veasley Jordan and Charles Thessel Jordan. Both parents had a high-school education; her mother was a homemaker, and her father was a meat cutter. At the time of her birth, both parents were well positioned in the Baptist Church. Lynda attended a racially diverse elementary school in Boston.

Lynda's parents divorced during her early years and, as the oldest child, it became her responsibility to help her mother run the household. After the divorce, Lynda, her mother, and two sisters moved to the Bromley-Heath Housing Project. At that time, Bromley-Heath was one of the worst public housing complexes in Boston. Approximately 85 percent of the tenants were on public assistance, as were Lynda and her family.

Lynda's mother remarried when she was eleven years old. Her new stepfather brought twelve children to the family, causing an exponential growth to the family structure, and causing Lynda to feel lost. The blended family moved to Dorchester, Massachusetts, where she attended Dorchester High School. Although the high school was predominately black, it was a part of Boston's integration strategy. The dropout rate was twice as high as the city's average; it was one of the poorest public school systems in the nation.[75] In 1971 Lynda, a sophomore in high school, was on the verge of being one of those high-school dropouts; she spent most of her time cutting classes and smoking cigarettes. Lynda's destiny changed on the day she ran into the high school auditorium to avoid getting caught for smoking cigarettes in the girl's gym locker room. She sat down in the midst of an assembly of a small crowd of students. She would later learn that these students were hand chosen by the teachers and guidance counselors as students whom they thought of as being "college material." She had no idea that there would not be an assembly going on at the time; nor did she know that there was an opportunity available for her, in this high school, to go to college.

As she sat down beside a student in the meeting, she heard Dr. Joseph Warren,[76] director of the Brandeis University Upward Bound Program, say, "What are you going to do for the rest of your life, stand on the corner of Blue Hill Avenue and smoke cigarettes?"[77] Jordan was stunned! It was as if this man, whom she had never met or seen before, was speaking directly to her! At the very moment that he spoke those words, she was sneaking into the auditorium trying not to get caught for smoking cigarettes. She said, "It really struck me, because that's exactly what I was doing—hanging out on the street corner smoking cigarettes. I had seen enough people do that to know that it would get me nowhere."[78] She was mesmerized and Dr. Warren had her attention. She sat listening to what he said about the Upward Bound Program at

Brandeis University.[79] At the end of his presentation, Lynda told Dr. Warren of her desire to go to college.[80]

Lynda was not offered the opportunity to be among the group of students in the auditorium who were given the chance to attend college. Until that time, she had been labeled by some of her teachers as someone who would not be able to further her education. But she applied for the Upward Bound Program was accepted, and entered the program during the summer after her sophomore year, and the following two summers. While at Brandeis, she spent six weeks with 100 other students in a place that seemed to be far away from the inner city. "It was only 54 miles from Boston, but it seemed like an eternity because there were trees and grass," she later said.[81]

It was during the first summer at the Upward Bound Program that Lynda became aware of her ability to do math and science. Halfway through the six-week program, Dr. Warren organized an Honor's Convocation to acknowledge the students who made the greatest academic improvement during the first three weeks of this vigorous program. To her surprise, Lynda was one of the six students.[82] The acknowledgment of her accomplishment at this assembly was her first hint that she really did have academic ability, and she now understood why Dr. Warren was working them so hard. She would gain self-confidence, as her determination grew; no one would be able to take that away from her. She worked so hard in the program over the next two summers that she surpassed the science curriculum that Dr. Warren designed for the program. In an effort to support Lynda's scientific development, Dr. Warren hired a private tutor to teach her organic chemistry during the summer after her junior year in high school. She graduated from high school in 1974.

It was Dr. Warren who encouraged her to attend his alma mater, North Carolina Agricultural & Technical State University (NC A&T). He knew that she would thrive there with people of the same background and culture. About her choice of college, she says, "I had to get out of the city—everyone I knew who stayed in Boston did not make it through college. I also knew I needed to be among strong black women."[83] There were plenty of role models for her at NC A&T. She arrived on campus with a suitcase held together with rope and an acceptance letter, but no housing. She also did not have financial aid to attend college. She managed to piece together funding through grants and work-study, a $500 Edwards Scholarship from the Edward's Scholarship Fund in Boston. At first she started to major in nursing because she felt she could help people and the field involved chemistry. During her chemistry course, she was influenced by the chemistry professor to change her major to chemistry. At the time, there were three women chemistry faculty members whom she admired, Dr. Jothi Kumar, Dr. Etta Gravely and Dr. Vallie Guthrie. She thrived on campus, majoring in chemistry, but she also engaged in the social scene. She became a member of Delta Sigma Theta Sorority and was president of the Beta Kappa Chi Scientific Honor Society. She graduated from NC A&T in 1978.

She went on to study for a master's degree in chemistry at Atlanta University (now Clark Atlanta) at the suggestion of the chairman of the chemistry department, Dr. Walter G. Wright. She had wanted to go to medical school, but changed her mind when she went to the Harvard University Health Careers Summer Program. She did not like the clinical part of medicine, which she had to observe as part of the course, but fell head over heels in love with biochemistry—a love affair that she still holds today.

Her master's thesis research was done under the supervision of Dr. James Story and Dr. Cyril L. Moore, at Morehouse Medical School. At Atlanta University, a component of the graduate student training in chemistry included the periodic administration of the Graduate Record Examination (GRE), so that the faculty was aware of how their students ranked at the national level. Jordan aced the exam, and the Massachusetts Institute of Technology (MIT) recruited her to pursue a PhD degree in biological chemistry. She graduated from Atlanta University with a master's degree in chemistry in December 1980. However, the summer prior to attending MIT, she worked as a summer intern at the Polaroid Corporation in Waltham, Massachusetts. She began her studies in the PhD program in biological chemistry at the Massachusetts Institute of Technology in September 1981. She said that at MIT she could enjoy full immersion in her work; she enjoyed being a "nerd among other nerds."

During the time that Lynda matriculated at MIT, most of the students were males, and only about 3 percent were African American. MIT was known for attracting high achieving minority students, but it was a chilly place for her. When her notebook consisting of research data and notes mysteriously disappeared, she was going to quit, but was intercepted by a fellow NC A&T State University alumnus, Dr. Ronald McNair. Dr. McNair was being honored by MIT and was on MIT's campus to give a talk. Jordan, who was the co-chairperson of the Black Graduate Student Association, was invited to go and meet him. Dr. McNair greeted Lynda in the reception line and held her hand to quote her statistics back to her. He knew she was an alumna of NC A&T State University, and he also knew where she was in her studies at MIT. When she recounted her dismay and desire to quit the program because of the disappearing research data, he told her to do the work over, and keep copies of the data at home. This advice from such a formidable role model and ally came at a critical time in her career. She continued her studies at MIT and earned a PhD degree.

Lynda Jordan was described by her professors as one of the brightest, most intelligent, and hard-working students to come through the PhD program. Her thesis advisors were Dr. Christopher T. Walsh and Dr. William Orme-Johnson. Her dissertation title is "Purification and Characterization of the Methyl Viologen Reducing Hydrogenase and a Flavoprotein from Methanobacterium thermoautotrophicum, Δ H." She received her PhD degree in 1985, becoming the third black woman of African descent to receive a PhD in chemistry from MIT.[84]

While in search of a laboratory to conduct postdoctoral studies, she wrote to several scientists in France who were conducting biochemical research. Several invitations were extended from different laboratories, but she chose to work with Dr. Françoise Russo-Marie because of the medical implications of her work. Lynda received a small grant from the French government, entitled "Centre de la École Superior," to study in France and would subsequently receive fellowships from the INSERM (Institut National de la Santé et de la Recherche Médical; the French Institute of Health and Medical Research), the Pasteur Foundation, and the prestigious Chateaubriand Fellowship. It was her desire to explore the world after her PhD and again get out of Boston, so she sought a postdoctoral position in Paris. She was treated very well in Paris. Françoise Russo-Marie was working on lipocortin, the protein inhibitor of phospholipase A-2 from non-human sources.

Dr. Jordan was able to extract the phospholipase A-2 enzyme from the human placenta. This enzyme has been linked to asthma, arthritis, pre-term labor, and a host of inflammatory diseases. Understanding the physiological basis of the enzyme enables one to identify the underlying cause of things like hypertension or kidney failure. The Europeans treated her well as a scientist and respected her despite her race, a general feeling of other blacks who had been working in Europe. Her fellowship lasted two years.

Upon her return to the United States in 1987, she was recruited by NC A&T as an assistant professor of chemistry. She did not want to go back to the chilly climate of MIT and accepted the position at NC A&T. The labs there were outdated and were not equipped to do her research. She started her research but had to go to the labs in Research Triangle in order to test her experiments. She was able to transform the chemistry department by writing grants to build a $2 million state-of-the-art biochemistry lab for her research and student education. Here she continued her research on PLA-2 and trained students. This was difficult because she had to train undergraduate students to isolate the PLA-2 from the other proteins that are in the placenta. The students had to follow a strict protocol or the protein would die. She enjoyed being at a historically black college because of the students. "Some people hear she's hard and tough, and they won't take her class for that reason," says Stephen Webb, one of her former graduate students.[85] Dr. Jordan knows that science is not a nine-to-five job and that you must be there to do the work. She is very proud of the accomplishments of all of her students. Three of her undergraduate students have received PhD degrees in biochemistry (two females) and chemistry (one male); four of her undergraduate students have received Master of Science degrees in chemistry under her tutelage (two males, one female). Two of her undergraduate students have received master's degrees at other institutions or programs (two females). One of Dr. Jordan's master's degree students also earned the Juris Doctor degree, and a host of her other students are now science educators.

In 1997 she accepted an appointment as one of the Martin Luther King Visiting Professors at MIT.[86] This appointment was to teach and conduct research for two years. During that time, she accepted the call to ministry, and after going back to NC A&T for one year, she returned to Cambridge, Massachusetts, to matriculate at Harvard University. Dr. Jordan is the first person in the history of Harvard University to *simultaneously* earn both the master of divinity and the master of public health degrees from Harvard Divinity School and Harvard School of Public Health, respectively. She conducted several ethnographic studies concerning poignant issues related to the black community, and she has written a paper entitled "Domestic Violence in the African American Community: The Role of the Black Church." She is also the founder, CEO, and senior pastor of a Place to Heal Ministries, Inc., in Cambridge, Massachusetts. Dr. Jordan is currently designing research studies that utilize her cumulative education to address the issues related to science and religion as related to spirituality and health.

CHAPTER 6

Industry and Government Labs

ESTHER A. H. HOPKINS

Dr. Hopkins is one of the few American women to have held a doctorate in science and a license to practice before the U.S. Patent and Trademark Office.[1] Her career included academia, industry, and government.

Esther was born Esther Arvilla Harrison on September 16, 1926, in Stamford, Connecticut. She was the second of three children born to George Burgess Harrison and Esther Small Harrison. Her father was a chauffeur and sexton at a church, and her mother worked in domestic service. Neither of her parents had an advanced education. Her father had some high school education; her mother attended only primary school. However, both of her parents wanted to make sure their children had a good education. When Esther was three and a half years old, her mother took her along to register her older brother for school. Because Esther was taller than her brother, the teacher suggested that she take the test to start school. She passed the test and was able to start kindergarten at the age of three and a half! She and her brother went to school together all through elementary school. Boys and girls were separated in junior high school; in high school they remained separate but attended the same school.

She decided in junior high school that she wanted to be a brain surgeon. This was because she met a woman doctor in Stamford who had an office in one of the buildings that her father cleaned. The woman was a physician and graduate of Boston University Medical School. Esther decided that she wanted to be just like her. Therefore, when Esther entered high school, she chose the college preparatory math and science track. She took as many science courses as possible in order to get into Boston University. She spent a lot of time at the local YWCA, becoming a volunteer youth leader. One speaker at a YWCA luncheon discouraged her from entering science and suggested that she become a hairdresser. Esther was hurt but not discouraged by this. She graduated from Stamford High School in 1943.

Hopkins's parents could only afford to send one of their children to college, and because her older brother was not interested in college, Esther became the lucky one to attend. Esther applied to the Boston University College of Arts and Sciences program and was accepted. When she arrived at Boston University in the fall of 1943, she was told by the dean of the college that she could not stay in the dormitories because they were "full."[2] Therefore, Esther had to find off-campus housing.

She decided to major in chemistry rather than biology because, at the time, chemistry involved thinking and reasoning, and biology was just rote memorization. She graduated from Boston University in 1947 and applied to medical school.

However, the post–World War II admissions to medical schools were reserved primarily for men and individuals with master's degrees, and many slots for men were reserved for World War II veterans. She was very disappointed when she did not get in and she cried for a week. A woman graduate student said to her "So, you didn't get in; what are you going to do with yourself?"[3] Well, having heard that, she picked herself up, stopped crying, and applied to graduate school to major in chemistry. Esther was accepted at Howard University, where she joined the research group of Lloyd Ferguson, a well-known African American physical organic chemist. She was his first graduate student and finished her MS degree from Howard in 1949.

Her first chemistry position after graduate school was in physiology at Howard University Medical School. It was here that she discovered her preference for laboratory work rather than interacting with people in the clinic, which she would have had to do as a medical school graduate. She preferred to be in the laboratory with the inanimate objects. In the lab she was required to prepare electrolyte solutions to administer to people. She wore a lab coat, so people (probably mostly black) did not know that she was not a doctor. She realized, at this time before modern medicines, that there was very little that a doctor could do to cure patients, and yet they thought she was "God." This drained her emotionally, as she wanted to help them.[4]

Her next job in 1949 was at Virginia State College in Petersburg, Virginia, which was a historically black college at the time. Having grown up in New England, the transition to Southern life was difficult. Her first experience with an all-black student body had been at Howard University.

Although she liked being in this all-black environment, especially being among black women whom she could mentor, she left this position in 1952.

On December 27, 1952, she married a Liberian man, John Payne Mitchell, whom she had met at Boston University, and they had one daughter, Susan Weamah Emma. They disagreed over the roles of women within the family; he wanted her to be a traditional homemaker. This disagreement, combined with his desire to return to Africa, led to their divorce.

From 1955 to 1959 Hopkins worked as a research chemist at the New England Institute for Medical Research in Ridgefield, Connecticut. She went to work for American Cyanamid from 1959 to 1961, about the time when corporate America began to look for qualified black chemists.[5] While American Cyanamid employed her, she realized that she would not be able to move up the corporate ladder with only a master's degree in chemistry. Companies did not want to hire her at the MS level because they would have to pay her more, but they also did not want to hire her at the PhD level unless she had the degree.[6]

On January 20, 1959, she married her second husband, the Reverend Thomas Ewell Hopkins. They had one son, Thomas Ewell Hopkins, Jr., and later she started to take stock of her life and career choice. As she says, "With my husband as sympathetic supporter, I looked at what I was, where I was, and what I would probably be doing for the rest of my life and decided that I would work. There was no life of leisure or moneyed indulgence on the horizon; so I decided to stop dilly-dallying and be a chemist. This required the union card, the PhD, so I set out to earn it in order to give validity to my abilities."[7]

She and her husband made the decision that she should pursue the PhD degree at Yale. Therefore, she applied to Yale and having passed all the entrance exams, she received a traineeship (tuition plus a small stipend for her first year of study from the United States Public Health Service) and worked for Dr. Jui H. Wang. Trying to manage her laboratory research and her children, a daughter in the sixth grade and a son in nursery school, she enrolled her son in the Yale University School of Theology nursery school on campus. Her son grew up thinking that fathers went to work and mothers went to graduate school.

There were eight women in the graduate program before Yale undergraduate school became coed. Work in graduate school was long and hard, but fulfilling. She enjoyed exercising her mind as some enjoy exercising their bodies. Her graduate research was in biophysical chemistry. She was interested in a chemical model for an active site for an enzyme and the kinetics of a model enzymatic reaction. The title of her thesis is "Catalysis of Phosphoryl Group Transfer by Metal Ions." She received her PhD from Yale in 1967, and her whole family came to the celebration. The migrant girl from South Carolina saw her daughter receive academia's highest degree from one of America's Ivy League universities. The dedication page of her thesis reads, "To the three generations of my family whose dreams and sacrifices are inextricably woven into this degree and all that it means."[8]

Since she was now armed with her PhD, she had to make a career choice: industry or a hospital in New Haven. Polaroid Corporation was interviewing Yale graduate students; it was the time when corporate America was recruiting black professionals as a result of the Civil Rights Act of 1964. A vice president of Polaroid was familiar with her work, and she was a black PhD woman willing to work for them; so it did not take much for them to invite her to join them. She says of Polaroid, "I found this strange company with an *esprit de corps* that was humanized and that urged people to look at inventive things they could do. While it was certainly not perfect, the company tried different ways of providing a work life for people which allowed them to continue to grow."[9] She worked for them for twenty-two years, working in the analytical lab and supervising a group in the emulsion lab, where they did photographic coatings. She had the occasion to talk with an engineer with whom she had interacted on the Polaroid–South Africa matter and found out that he was currently going to law school. She heard herself say, "Were I to start over, I'd be a lawyer." This shocked her, and she realized she could not start over but she could do something; if she really wanted law, she should try for it.[10]

She worked at Polaroid during the day and began attending night classes at Suffolk Law School in 1974. While studying law, she applied for and received a career exposure experience in the patent department of Polaroid. It took her three years to get the degree, and she persevered in spite of being injured in an automobile accident. She did not want to lose momentum. In 1977, she received her JD degree from Suffolk and spent her last eleven years at Polaroid in the Patent Department until her retirement in 1989. She loved patent law because she was able to combine her scientific knowledge with the legal terms to write the patents. Her description of patent law suggests that a chemist who has made a novel compound comes to her to explain his or her new compound. Hopkins, as the patent attorney, takes this description down in a logical, scientific, legal, and candid form by interacting personally with the chemist to describe the new compound in legal terms. Then, having written the document (patent), the patent attorney presents the invention to the U.S. Patent and Trademark Office and acts as its advocate, pointing out where it meets society's requirement for its seventeen-year protection.[11]

Despite her retirement from Polaroid in 1988, she was not ready to stay home and live a life of leisure. Within a year after her retirement, she became deputy general counsel at the Massachusetts Department of Environmental Protection. She was the fiscal and administrative counsel, responsible for contracts, grants, and trusts. She advised on conflict of interest issues and handled personnel matters. In this job, her work was to protect the environment. She said, "Environmental protection—preserving air and water—is as vital and fascinating as patent work. The central question is this: How can we give our children a clean world? One way of working to do so is to organize, regulate, and respond to emergencies. 'Clean-up' is a very important part of it."[12]

Dr. Hopkins ran for selectman of her town of Framingham, Massachusetts, the first time unsuccessfully, but won in 1999, becoming the first African American selectman of the town.[13] In 2002, she became the chair of the board of selectmen, essentially the mayor of the town of Framingham.

You would think that, with all of this work and study, she had no time for relaxation and volunteer work. Quite the contrary, she was active in religion and music. She played the organ at church while working at medical school and sang in the Choral Arts Society as an undergraduate.

She ran a choir for men and boys at Petersburg when she was at Virginia State College. She currently sings in the choir, plays the organ, and sometimes preaches sermons at her church. Her husband was a social worker and Unitarian Universalist Minister.

She was a trustee at Boston University for twenty-two years and is now a trustee emeritus and an overseer. She is also active in the alumni association of Boston University. In 1995, she received the Boston University General Alumni Association Award for Distinguished Service to Alma Mater. All this is ironic, considering that when she was admitted she could not stay in the dorms. She was honored for all her volunteer work, which is too numerous to mention by the Boston University Alumni Association.

She is very active in the local and national American Chemical Society (ACS), serving on boards and committees. She is a trustee of the Northeastern local section of the ACS. In 2010, she received an award for being a sixty-year member of the ACS.

She was featured in a poster that Nabisco published on "Famous Black American Scientists." This poster was widely distributed so that she received letters from a fourth-grade class at a science and music magnet school in Houston, Texas. One of the letters read:

Dear Mrs. Hopkins,
When I grow up, I hope to be just like you, have a good education, study a lot, and maybe get two degrees. We learned that if we try hard we might be able to accomplish what you have done. I hope you come and visit us some time.
Love,
Desiree Dorusseau[14]

She was very delighted and pleased to be considered a role model by this young child and by African American women and men of any age.

BETTY WRIGHT HARRIS

Betty Harris is known for her invention of a spot test for explosives 1,3,5-Triamino-2,4,6-Trinitrobenzene (TATB) that is now used by the U.S. Department of Homeland Security (DHS) to screen airline passengers to determine if nitro aromatic explosives are present.

Betty was born on July 29, 1940, on a farm in northeastern Louisiana, the seventh of twelve children born to Henry Hudson "Jake" Wright and Legertha Evelyn Thompson Wright.[15] She lived on a plantation with her family until the age of twelve. They were sharecroppers whose owner charged them one-quarter of their profits. However, they were successful farmers, raising cows, horses, hogs, chickens, and many crops such as cotton, corn, sugarcane, peanuts, potatoes, soybeans, and other vegetables. They also had fruit trees around the house: figs, peaches, pears, and apples. The Wright children learned much from the farm industry, including the value of a family working together. They had respect for the land, persistence, hard work, and hope for the future.[16] Her mother taught school, and both parents encouraged their children to get an education or learn a trade.

When Betty was twelve, her father paid cash for some farmland and built a new four-bedroom house that sat on the border of two school districts, Ouachita and Caldwell Parishes. She could have attended either but chose

Union Central High School at Caldwell Parish (Columbia, Louisiana) with W. P. Moore as principal. This school had an excellent curriculum that was based upon a rotating block system. Their teachers were African American and most had master's degrees. All were dedicated to excellence in education.[17] She was encouraged to pursue science as a career.[18]

Living on a farm, she missed many months of school because of the labor required for the planting, working, and harvesting of good crops. Her substitute English teacher assigned her the Complete Literature book. Betty read and wrote a synopsis of each item, which she later discussed with the teacher. This experience helped her to develop a passion and appreciation for literature.

In 1957 she entered Southern University, a historically black college in Baton Rouge, Louisiana, and majored in chemistry with a minor in mathematics. Dr. Jack H. Jefferson, department chairman, and Ms. Celestine Tillman were her mentors. Betty graduated from Southern University in January 1961 at the age of nineteen. In July 1960 she married Alloyd A. Harris, also a Southern University graduate; they later had three children.

In 1961 she received a research assistantship at Atlanta University, where she studied for two years and obtained an MS degree. Dr. Kim Huggins was her research advisor and Dr. Henry McBay[19] was her course and seminar advisor. She graduated from Atlanta University in August 1963. Her thesis was titled "Studies of Conjugated Systems: Some Addition Reactions of 2,3-DiMethyl-1-(4-Nitrophenyl)-1,3-Butadiene."

In September 1963 she became assistant professor of chemistry and mathematics at Mississippi Valley State

College, now Mississippi Valley State University (MVSU), which is a historically black college at Itta Bena, Mississippi.[20] She remained at MVSU for only one year because the college did not have the budget and living facilities to accommodate its faculty. She then accepted an assistant professor position at Southern University in New Orleans (SUNO), Louisiana, where Dr. Paul T. Groves was the department chairman. She stayed at SUNO for ten years. In 1966 she took an educational leave of absence from SUNO and accepted a graduate assistantship in chemistry at the University of Wisconsin at Milwaukee, where she remained for two years until she completed her course work for the doctors of science degree. Around 1968 she accepted a summer internship at IBM at Fort Collins, Colorado. She was asked to run the atomic absorption spectrometer.

FIGURE 6.1 Betty Wright Harris Credit: Photo courtesy of Los Alamos National Laboratory

She said that she "had never seen one, not to mention, run one!"[21] However, she and her husband packed up the kids and moved to Boulder, Colorado, for summer work, she at IBM and he at Ball Brothers.

After IBM, Dr. Harris returned to SUNO but continued to seek new career opportunities. She revisited her acceptance at Los Alamos Scientific Laboratory (LASL), now Los Alamos National Laboratory (LANL), which she had received at the same time as the offer made by IBM. In the summer of 1970, she became a visiting staff member at Group WX-2 of LASL. The Q clearance from the United States Department of Energy (DOE) was still valid and she held it for forty years. It permitted her access to secret restricted data (SRD) needed in her research projects.

After the first summer at LANL, she was selected to participate in the Science Technology and Engineering Program (STEP) that provided additional technical training for potential full-time employees. She became a member of the high explosive (HE) synthesis and development team. It was through them that she completed the research for the PhD and began writing her dissertation. Also she was required to complete her residency at the University of California or the University of New Mexico (UNM). At the time, she had four children, three of her own and her nephew. The two younger ones were living with their grandmother. She decided that moving to California was too expensive and that she would be too far removed from her advisors, Dr. E. P. Papadopolous of UNM and Dr. Michael D. Coburn of LANL. She chose to do her residency at UNM, where she received a one-year teaching assistantship. During this time, she taught undergraduate laboratory classes and continued to write her dissertation, which included "Reactions of 2-Aminopyrimidine with Picryl Halides."

In January 1973, she was excited about completing the PhD requirement and returning to SUNO, only to learn that this was not advisable because of the chemistry department's mid-year budget. Instead, she accepted a one-year assignment as an assistant professor of chemistry at Colorado College in Colorado Springs, Colorado. This was a wonderful experience. During this time, her advisors found a few items in her dissertation with incomplete data. She was allowed to finish these experiments at the Air Force Academy's Frank J. Seiler Research Laboratory under the oversight of Dr. Raymond McGuire. By December 1973, she received her PhD in chemistry from UNM. The position at Los Alamos became permanent. She left Colorado College and returned to LANL and found her laboratory and safety shoes waiting. Ruben Sandavol, the property clerk, handed the latter back to her and said, "I knew you would be back."

Work at LANL was exciting and enjoyable. There was a technical staff with a wealth of knowledge and instrumentation that they readily shared. The opportunities to learn more and expand her expertise increased. While still with the high explosive research and development team, she became the group M-1 (GMX-2) safety officer and led the DOE Tiger Team's required environment cleanup of the LANL explosive corridor. This involved identifying and determining the concentration and sometimes the movement of hazardous chemicals and materials from research laboratories, decommissioned and active buildings, test sites, firing sites, and disposal sites. Effluent from process and development building was also sampled and analyzed, as was the surrounding soil. This was the subject of many research projects for her students.

She then became responsible for more challenging projects as her expertise increased. She had seven years of managing research projects in explosives and analytical and organic chemistry method development. Her other experience includes three years of corporate management as chief of chemical technology for a gas turbine company with budget responsibility and one year as acting director for LANL student internship programs. She is an expert in the chemistry of explosives. Dr. Harris's major projects at Los Alamos Laboratory have included the development of a spot test for identifying explosives in a field environment for which she holds the patent that was issued in 1986. The Department of Homeland Security now uses this spot test when people are checked in at airport security and other places.

Dr. Harris is interested in getting more women and minorities to become scientists. She developed a summer church project at Los Alamos for high school students, encouraging them to prepare for the SAT and ACT tests. She did this with the help of their parents who worked at LANL. She is very proud of this accomplishment because most of the students succeeded in getting into college and going into professions. At Los Alamos she was assigned to the laboratory's Diversity Office, where she worked on a program to better utilize the résumés that the Human Resources Division receives from underrepresented groups. She also worked on a project that would prepare members of an underrepresented group for positions in upper management.[22]

Dr. Harris was glad that she switched from working in academia to working in a government research laboratory like Los Alamos. She is quoted in an article in *U.S. Black Engineers and Information Technology* as saying that she "was fortunate to have at her fingertips, a lot of resources, including one fantastic library, money to buy equipment that is needed and work with the best experts in the field."[23] When she worked in research laboratories of HBCUs, they were not as well equipped as the government laboratories because of budget problems.

Dr. Harris was actively involved in science education outreach programs in northern New Mexico, acted as a mentor to college students working summers at Los Alamos, and was responsible for the development of the Girl Scout Chemistry Badge, which is similar to the Boy Scout Merit Badge. Her other activities include serving as president of the New Mexico Business and Professional Women's organization and twice as chair of the New Mexico Section of the American Chemical Society. In 1999, she received the Governor's Award for Outstanding New Mexico Woman. In 1996, Harris was one of eight women profiled in a CD ROM entitled "Telling Our Stories: Women in Science."[24] It was for girls aged ten and up and contained a profile of her life and an activity that she helped design.

Career successes for Dr. Harris held some trying and painful times. In the late 1960s her husband studied and received his MS degree in mathematics at Atlanta University. They were separated in 1970 and later divorced in 1972. She is the mother of three children, with five grandchildren, and three great-grandchildren.

She retired from LANL in 2002 and became a DOE trained and certified document reviewer, working for several small business contractors: Excalibur Associates, Columbia Services Group, and SMARTEC. Her assignments were with the Office of Classification at DOE headquarters in Germantown, Maryland, and the National Archives and Records Administration II (NARA II) at College Park, Maryland. This office determines which documents should remain classified and which can be released to the public. She could do this because she still had a Q clearance, which meant that she could review documents containing Secret Restricted Data.

Her advice for young people who might be interested in science is, "I say, do it. It is worth the effort. The pursuit will sometimes be difficult. You may not have the money, the time, the basic life necessities, or even the results on your experiments that advisors want. But, get the degree(s). For you, the world will open up. The view, the opportunities, the salary will all be beautiful."[25]

Personal Closing Note from Dr. Harris

I thank Dr. Edward S. Jenkins, my high school science teacher, for two projects that made me think, write, and wonder about the possibilities of science:

- Project 1 was making firecrackers from basic materials
- Project 2 involved fermentation and counting yeast cells under a microscope

The product from Project 1 did not explode with a loud bang and had to be redone. Project 2 did explode and I had to clean up my mother's pantry where I was keeping my jars in which I grew the yeast. I forgot to open them at the interval given in the experiment.[26]

SINAH ESTELLE KELLEY

Sinah Kelley made history in the African American community by being a vital part of the U.S. Department of Agriculture team that worked on the mass production of the antibiotic penicillin.[27]

Sinah Estelle Kelley was born in New York City on April 23, 1916. She was the daughter of William Melvin Kelley, Sr., and Gladys Caution Kelley. She had one younger brother, William Melvin Kelley, Jr., born in 1937. Her father was a newspaper editor; information about her mother's career was unavailable. Sinah was educated in a private school, the Ethical Culture School of New York City, until the eighth grade. She then went on to the affiliated Fieldston High School in Riverdale.[28] Very active in science and sports (basketball) while in school, she graduated from the Fieldston High School in 1934.[29]

She enrolled in Radcliffe College (now a division of Harvard University).[30] When she arrived at Radcliffe, she found out that she was one of two black students who would be in the class of 1938.[31] Sinah commuted to Radcliffe and did not live in the dorm.[32] She majored in chemistry, and one of her professors was Dr. Louis Feiser, a well-known organic chemist; she used his name as a reference when she applied for a job.[33] In addition to her chemistry studies, her extracurricular activities were science club, varsity basketball, and the choral society. During the summers of 1937 and 1938, she was an assistant laboratory technician in the Department of Pathology in Harlem Hospital under Dr. Solomon Weintraub, a pathologist, doing routine and special chemical analyses on blood and other biological fluids.[34] She graduated from Radcliffe, receiving an AB in chemistry, in 1938.[35]

Kelley began her career by continuing to work at her Harlem Hospital summer job in the Department of Pathology and in the Pneumonia Research Laboratory, where she worked for Dr. Jesse G. M. Blullowa, who was the director of that lab.[36] In 1940 to 1941, she took two graduate courses at New York University (NYU), Educational Psychology and the Teaching of Science (Chemistry).[37] She then decided to apply for a job in one of the federal research laboratories. These labs were open for minority chemists during World War II.[38] In her biography at the Tenth Annual Radcliffe Reunion, Kelley states that she first went to Southern Indiana in 1942 to work in a War Department laboratory testing smokeless powder. When that lab closed, she transferred to a lab in Pennsylvania to work on tri nitrotoluene (TNT). In 1944, she transferred to the U.S. Department of Agriculture Northern Regional Lab in Peoria, Illinois, working in the Fermentation Division. It was there that she worked on the mass production of penicillin.[39] Her work consisted of the analysis of the products produced by the microorganisms in the termination process. She has written about fermentation products in several publications.[40]

Kelley did not like living in the Midwest in Peoria but she adapted to life there. She preferred the East Coast, especially New York City. She had a garden plot on the grounds of the lab and played some basketball, although she said that age was catching up to her.[41] She also took some courses related to her work, including advanced organic chemistry. She remained in Peoria until about 1958, when she returned to the East to work in an Atomic Energy

FIGURE 6.2 Sinah Estelle Kelley Credit: Photo courtesy of MARBL Library Emory University

Lab on Hudson Street in New York City, as stated in the Radcliffe College Twenty-fifth Anniversary Reunion booklet. She worked on special research tracing the effects of strontium 90.[42]

Kelley received a letter of dismissal from the Atomic Energy Lab for poor performance in March 1972.[43] She fought back because at that time she had worked in federal labs for thirty years. She stated that the reason that she seemed to be slow was that work on the analysis of samples for stable strontium by flame photometry had not been done by anyone in the lab before and she had to develop the technique. She also had ordered a new instrument and had to get the bugs out of this new instrument before she could make it work. This information appeared in a letter to her congressional representative, Charles Rangel.[44] Congressman Rangel did take action by writing to the Atomic Energy Commission on her behalf. The outcome of this could not be determined, but Kelley retired sometime thereafter, in the 1970s.[45]

Kelley never married but devoted her life to the causes that she loved. She loved jazz music by Duke Ellington and wanted it to be played at her funeral. She was an avid reader of newspapers and books and even wrote a letter to the editor to the *New York Times* about racism. This was rejected twice by the editors of the *Times*, not because of its content, but because it was too long.[46] She loved tennis and would watch it on television. In her retirement she continued her interest in the life of her beloved Harlem and its citizens. She supported the League of Women Voters, American Association of University Women (Peoria Branch), Symphony of the New World, Harlem Consumer Cooperative, Schomburg Library, and the Dance Theatre of Harlem. She was also a member of the American Chemical Society.[47]

Sinah Kelley died on December 21, 1982. The memorial service for her was held at the Ethical Cultural Society on February 7, 1983, where her friends memorialized her.

KATHERYN EMANUEL LAWSON

Katheryn Lawson was one the first African American women chemists to be employed by Sandia National Labs in Albuquerque, New Mexico.

Born Katheryn Emanuel on September 15, 1926, in Shreveport, Louisiana, to John Venus Emanuel and

Ida L. Gillispe, she may have been their only child.[48] She attended racially segregated schools. She graduated as the salutatorian of her high school class in 1941. Katheryn then attended Dillard University in New Orleans, a historically black college,[49] where she earned a BA *cum laude* in the natural sciences in 1945. Next, she went on to study for an MS degree in organic chemistry from Tuskegee Institute in 1947. During the period between 1947 and 1951, she held a series of temporary appointments at historically black colleges as an assistant professor of chemistry, working at Bishop, Savannah State, Talladega, and Grambling. From 1951 to 1954, she advanced from assistant professor to associate professor of chemistry at Central State College. It is likely that she continued in academic positions at historically black colleges because she could teach with an MS degree.

The Cold War and the space race led to increased scientific funding in the United States as well as more graduate studentships.[50] Because of this new federal funding, universities were reaching out to U.S. students with offers of assistantships. Therefore, in 1954 Katheryn Emanuel received an offer of an assistantship to study for a PhD from the University of New Mexico.

Two months before beginning her new position in New Mexico, she married Kenneth Lawson, a chemist-bacteriologist, in 1954. Her husband was also studying for his PhD at the University of New Mexico and accepted a position as a laboratory technician at the city's sewage plant. He received his PhD in 1955. Katheryn Emanuel Lawson completed her doctorate in physical chemistry with an emphasis on radiochemistry in 1957. Her thesis advisor was Dr. Milton Khan, and her dissertation is titled "Behavior of Indium at Tracer Concentrations."[51]

FIGURE 6.3 Katheryn Emanuel Lawson Credit: Photo courtesy of Sandia National Laboratories

In 1957–1958, she worked as a staff member in biochemistry at the Veterans Administration Hospital in Albuquerque, New Mexico. Then, in 1958, she was recruited by Sandia National Laboratories. She was hired as part of Sandia's initiative to recruit more PhD scientists to develop advanced weapon designs. Like most of those scientists, she worked in materials research, conducting spectroscopic studies analyzing the molecular structure of irradiated materials.[52] She worked in the Crystal Physics Research division of Sandia Corporation, which is an ordnance-engineering laboratory under contract to the Atomic Energy Commission.

Katheryn Lawson was the mother of two sons, William Lawson and Kenneth E. Lawson, Jr. Her oldest son was born while she was in graduate school. The Lawsons were featured in an *Ebony* magazine article in 1965 about their life in Albuquerque as two of the handful of black professionals who lived in the city.[53] They were in demand to give talks about their work. At the time, there were only four doctors, one accounting firm, and one elementary school superintendent in a black population of 6,000. Nevertheless, the Lawsons loved the town and its diverse Native American population.

Katheryn Lawson took part in the National Urban League's Black Executive Exchange Program, in which she had two interests: advising young blacks about the quality and extent of personal effort and commitment necessary to attain even a small portion of the American dream; and advising white managers that the best way to help blacks achieve advancement is simply by not being afraid to recommend them.[54]

Dr. Lawson has numerous awards and publications, including a chapter of a book *Infrared Absorption of Inorganic substances*.[55] Dr. Katheryn Lawson died in Michigan on September 25, 2008.

CHAPTER 7

From Academia to Board Room and Science Policy

REATHA CLARK KING

Reatha Clark King is a woman who began life in rural Georgia and rose to become a chemist, a college president, and vice president of a major corporate foundation.

Reatha Belle Clark was born in Pavo, Georgia, on April 11, 1938, the second of three daughters born to Willie and Ola Watts Clark Campbell. Her mother Ola had a third grade education and her father Willie was illiterate. Her families were sharecroppers in Pavo. Her mother and grandmother raised her in Moultrie, Georgia, after her parents separated when she was young. She and her sisters worked long hours in the cotton and tobacco field during the summer to raise money. She could pick 200 pounds of cotton a day and earn $6.00, which was more than her mother's salary as a maid.[1] In the 1940s in the rural segregated South, the only career aspirations for young black girls were to become a hairdresser, a teacher, or a nurse.[2]

Reatha started school at the age of four in the one-room schoolhouse at Mt. Zion Baptist Church. Still more than a decade before *Brown v. Board of Education*, Reatha's schools were segregated. The teacher, Miss Florence Frazier, became Reatha's first role model. Reatha said, "I never wondered if I could succeed in a subject. It was only a question of whether I wanted to study the subject."[3] She later attended the segregated Moutrie High School for Negro Youth. Despite missing much school to attend to fieldwork, Reatha maintained her studies. She graduated in 1954 as the valedictorian of her class.

Reatha received a scholarship to enter Clark College in September 1954, originally planning to major in home economics and teach in her local high school. These plans changed after her first chemistry course with Alfred Spriggs, the chemistry professor. He encouraged her to major in chemistry and go to graduate school. She found that chemistry was the perfect major for her. She says, "Both the subject matter and methodology were interesting and challenging; the laboratory and lecture sessions were exciting; and my fellow students in chemistry were both serious students and fun to work with."[4]

During the summer vacations at Clark College, Reatha worked as a maid for a woman in upstate New York, who encouraged Reatha to visit New York City on Thursdays, her day off from work. Through the generosity and mentorship of her employer, Reatha, a young woman from the farms of Georgia, grew accustomed to the hustle and bustle of the big city.[5]

In 1958, she graduated from Clark College with highest honors. She received a Woodrow Wilson Fellowship for graduate study in 1958–1960. She also received a grant

FIGURE 7.1 Reatha Clark King Credit: Lee Prohofsky Photography

from the State of Georgia because of the United States "Separate but Equal Rule" for higher education. Higher education in the state of Georgia remained segregated in 1958; however, the state had to give her a grant to attend graduate school in another state since she was qualified to attend the white state university in Georgia.[6]

At the University of Chicago, she received a master's degree in chemistry in 1960.[7] She was able to bypass the master's thesis because she was also pursuing the PhD degree. She then received a National Medical fellowship to continue her studies toward the PhD in chemistry, which she received in 1963. Her PhD research was in physical chemistry and thermochemistry. The title of her PhD thesis is "Contributions to the Thermochemistry of the

Laves Phases." She had to learn thermochemistry and thermodynamics. She says that the University of Chicago focused on the research process and thus she had very good training there. She had two publications of her research before she received her PhD. The University of Chicago seemed at this time to be the cradle for many African Americans to receive PhDs in chemistry.[8] At this time, there were few women in the graduate program so that at times she was lonely. Dr. King says about her years at the University of Chicago: "Those four and one half years at the University of Chicago were a marvelous introduction to the advanced studies in other physical sciences as well as chemistry. Even though specialized in chemistry my constant association there with students and faculty in physics, mathematics, computer science, and geology greatly enhanced my academic and intellectual development."[9]

Reatha Clark met her future husband, N. Judge King, Jr., while she was in graduate school. He was a native of Birmingham, Alabama, and a graduate of Morehouse College. They married in 1961 while she was at the University of Chicago pursuing her studies and he was a graduate student in Atlanta.

After receiving her degree, she began looking for jobs in the Washington, D.C., area near her husband, who would be studying for his PhD at Howard University. Chemistry positions in industry were just beginning to open up for African Americans due to the space race.[10] She interviewed with one major corporation, which was more concerned with her gender and the fear of her becoming pregnant than her race. However, the federal government had employed African American chemists before most corporations. Reatha Clark King received offers from the

Naval Research Laboratory and National Bureau of Standards; she chose the National Bureau of Standards because it was closer to where she wanted to live. She started work in January 1963 under Dr. George Armstrong, becoming the first African American woman chemist to work for that agency. In fact, Dr. Armstrong willingly interviewed and hired an African American because he was enlightened about the issue of race and had a progressive attitude toward African Americans.[11]

The work at the National Bureau of Standards proved rigorous and demanding. She was put to the test when the Maritime Administration sent a rush test for a fuel sample analysis. When she called in the results at 3 A.M., the person waiting yelled with joy. It seemed her results matched that of an identical fuel sample that had been sent earlier. They did not know it at the time, but the agency was in dispute with another agency and needed the results to prove they were right! The Bureau received laudatory letters for her work.[12] She also worked on a publication about the heat of formation of oxygen difluoride, which took three years to complete. She received an outstanding performance rating, a promotion, and, along with her supervisor, an outstanding publication award for her division. She developed new techniques for flame calorimetry, later used for studies on fluorine and chlorine trifluoride with hydrogen.

Her career as a research chemist was extremely productive; before the age of thirty-one, she had produced eight quality scientific publications. She says, "Professionally, these years at the Bureau were my formative years, the ones that have most strongly influenced my professional style in other jobs."[13] It was while working at the National Bureau of Standards that she had two sons, N. Judge King III (Jay), born in 1965, and Scott, born in 1968. She juggled work and family life with the help from her husband and a child-care worker. She saved her vacation so that she could take time off when the children became ill. There were maternity leave policies, of which she took advantage. However, she and her husband shared the child care so that they could be with their sons.

In 1968, she moved to New York City when her husband accepted a position as chair of the Department of Chemistry at Nassau Community College. She was fortunate to find a position as an assistant professor of chemistry at the newly formed City University of New York, York College. She advanced rapidly through the ranks, becoming associate dean for the Division of Natural Sciences and Mathematics in 1970. She advanced to full professor in 1974 at York College and managed to continue an active research program despite the demands on her time as associate dean for Academic Affairs.

While working at York College, she attended the Columbia University Business School in order to obtain an MBA. She funded her studies at Columbia with a fellowship from the Rockefeller Foundation. Her objectives in taking MBA courses were fulfilled because she learned human relations management and organizational structure, which she was to put to good use in the future. She also loved learning for learning's sake, which is what she advises students to do. She received her MBA from Columbia University in 1977.

In 1977, she became the second president of the six-year-old Metropolitan University in the twin cities of St. Paul and Minneapolis, Minnesota, thus, becoming one of the few African American women to become president of a major university. She defined her job as president of the tax-supported institution as providing strong leadership to firmly establish the institution within the community. Together with the other employees, they created strong administrative and student support services, new academic programs, and dramatically increased the student enrollment and graduates. The enrollment increased from 1,600 in 1977 to 6,000 in 1988, and graduates increased from 1,200 to 5,500. The university received numerous awards for the work her team accomplished through networking with individuals and organizations already established within the community. She reinforced a "value system and organizational culture" that everyone in the university appreciated. Her administration established the University Development office, which enabled the University to seek donor funds for programs and scholarships. The University currently has two scholarships in her name: the Reatha Clark King Endowed Scholarship and the N. Judge and Reatha Clark King Endowed Scholarship.

In 1988, she was approached by the CEO of General Mills to become vice president of General Mills and president and executive director of the General Mills Foundation. She was recruited because of her accomplishments in firmly establishing Metropolitan University. The General Mills Foundation was established in 1954 to make grants to tax-exempt organizations in support of positive civic efforts such as education, planned low-cost housing, health, and nutrition. As CEO, she directed $45 million charitable giving programs. Although she held this position for fourteen years before retiring in May 2002, she remained vice president of General Mills and also served as board chair of the Foundation until 2003.

In 2004, she was selected by the University of Minnesota's Hubert H. Humphrey Institute of Public Affairs to receive the second Louis W. Hill Jr. Fellowship. The one-year fellowship provided financial aid and administrative support to study important issues facing philanthropy. At the end of the year, there was a symposium in which she was to report her findings. Dr. King's expertise in both academia and philanthropy made her the perfect choice. Dr. J. Brian Atwood, dean of the Humphrey Institute, said, "This fellowship is a capstone to [Dr. King's] exemplary career in science, higher education and as a corporate and foundation president." Accepting the fellowship, Dr. King commented about the lack of scholarship on philanthropy, "Nationally, knowledge about the field is shared mostly by word of mouth. As a profession, we have not developed the more traditional methods of research, as in chemistry. This fellowship gives me the opportunity to study several issues important to the field."[14] Dr. King's report of her research is entitled "Philanthropy and Public Policy: Working Together to Make a Bigger Difference."

As of 2010, Dr. King is retired from General Mills but remains busy in corporate governance. She serves on the boards of Exxon Mobil Corporation and the National Association of Corporate Directors (NACD). She is a Life Trustee of the University of Chicago and a member of the Executive Leadership Council in Washington, D.C. She is a former corporate board member of Wells Fargo Company, H. B. Fuller, and Lenox Group, and a former member of the Boards of Trustees of Clark Atlanta University, the Congressional Black Caucus Foundation, and the International Trachoma Initiative. She has received numerous honors and awards, including fourteen honorary doctorate degrees.

MARGARET ELLEN MAYO TOLBERT

Margaret E. M. Tolbert is a Renaissance woman. She has risen from her birth in rural Virginia to become a leader in the field of chemistry and academic, industrial, and government administration.

Margaret was the third of six children of J. Clifton and Martha Artis Mayo. She was born on November 24, 1943, in Suffolk, Virginia. Her mother was a domestic worker; her father served in the U.S. Army during World War II and became a landscape gardener upon his return home. When Margaret was young, her mother and father separated. Her mother tried to support all six children on her own but became ill and died. This left the children alone and, in order to keep the family together, the neighbors decided to take turns raising the children rather than call in the social service agency. This was the rural south and the people believed in looking out for their own children. However, this became difficult to do and the children were sent to live with their paternal grandmother, Fannie Mae Johnson Mayo. Since there were so many children, each one of them had to do chores around the house, and the focus on work rather than reading and writing at an early age caused Margaret to have problems early in kindergarten. However, when she entered first grade at Ida V. Easter Grade School, she began to realize that education was the key to her achieving a better life. This was the segregated South; therefore, she attended the school for black children in her neighborhood.

Her father died, and later her grandmother became ill and could not care for the family. Margaret had been working as a maid for a middle-class African American family who lived in another neighborhood of Suffolk. They decided to employ her as a full-time maid and nanny for their son, Alfred, while she attended high school. Earlier attempts by Mrs. Delia D. Cook to discuss official adoption of Margaret were met with angry remarks from her grandmother. Therfore when she graduated from Ida V. Easter Graded School at the end of the sixth grade, Margaret had to enroll in East Suffolk High School on the other side of town. She attended that school from grades seven through twelve; since there were no school buses, she had to walk two miles each way to attend school. Her oldest sister and her sister's husband took Margaret and her siblings in after it was clear that her grandmother was too ill to return home to care for them. In the meantime, Margaret had served as head of the household, taking care of her three younger siblings until her oldest sister assumed the responsibility. Margaret was not satisfied with living with her sister and her family, but she endured. It was during that period when she was a teenager that the Cook family virtually adopted her. They, as well as her high school principal, urged her to plan for college since they were college educated and recognized her potential in both science and mathematics.

The Cook family took her to Virginia State University to introduce her to some of the faculty members whom they knew there. The Cooks joined with teachers and the principal at East Suffolk High School in encouraging Margaret to take advanced placement courses in mathematics and science in preparation for college. The Cooks also took her to visit other colleges so that she could decide which college to attend. Margaret decided to attend Tuskegee Institute (now University) in Tuskegee, Alabama. She chose this university because it had a family atmosphere and her high school principal recommended it highly. Additionally, the Cook family had friends on the faculty there. Margaret also wanted to get away from Virginia to start a new life. She had received a full scholarship from two other universities, but only a partial scholarship (work-study) from Tuskegee. Mrs. Cook made several dresses and skirts for Margaret, and her brother made a beautiful purple coat for her to wear while enrolled at Tuskegee.

The Cooks drove her to Tuskegee and made sure that she was comfortable in her new environment. Before leaving Suffolk, churches in her hometown gave her money, and individuals in her neighborhood gave her gifts, which proved useful once she entered Tuskegee. They were very proud of her decision to attend college, and they wanted to help her in any way that they could. Members of the Tuskegee faculty helped her adjust to college life. They often invited her to dinner at their homes and counseled her on life and the educational process. Also, they encouraged her to complete her studies and earn her degree. People from Suffolk who visited Tuskegee gave her gifts of money and other items when they visited her on campus. Although she had a few dollars, she decided not to make any trips home on holidays after her freshman year due to the long bus ride, which took at least twenty-four hours.

Margaret first decided to study medicine when she was in high school. Her interest in medicine remained strong as she entered Tuskegee, but she soon realized that she would have no money for medical school. She found out that if she studied basic science there would be support for further study in graduate school, so she decided to study chemistry. Because she excelled in her studies, she was placed in the rigorous series of courses that are now advocated by the American Chemical Society for chemistry majors. Margaret became an undergraduate research assistant to Professors Courtney J. Smith and Lawrence F. Koons, faculty members of the Tuskegee Department of Chemistry. As a research assistant, her research problem was to study the capacity of different chemicals to conduct electricity in solutions under varying degrees of resistance. She learned that the difference in conductivity depends in part on the ease with which the molecules disperse in solution.

During the summers, she participated in summer internships at Central State University and Argonne National Laboratory. At Argonne, she was a member of the analytical group that studied the various chemical compounds of uranium samples; she therefore learned to work with radioactive compounds under supervision. She became a pioneer because she was the only woman in her class who completed the chemistry curriculum in 1967 and earned a BS degree in this field (with a minor in mathematics) from Tuskegee. Since Margaret's goal was to pursue a career in chemistry, she knew that she needed to complete graduate school in order to obtain a good position. Some of her advisers told her to go directly to the PhD, but because she did not know what the future might hold for an African American woman with a PhD in chemistry, she decided to earn an MS degree first.[15] She therefore entered Wayne State University in the fall of 1967 to pursue a degree in analytical chemistry. Wayne State University gave her funding to pursue her degree, which she obtained in December 1968.

Margaret Mayo decided to return to Tuskegee where she became a professional research assistant working for Dr. Lu and Dr. Chung in the School of Agriculture. She conducted research on the effects of different chemicals on chicken liver enzymes. This research was very different from the projects on which she had worked as an undergraduate student. She also served as a lecturer and laboratory instructor for the Tuskegee Institute Teachers' Training Summer Institute and later taught mathematics at Tuskegee.

In 1970, because of her experience in science and probably because of the need to increase the number of minority PhD students on its campus, Brown University, an Ivy League university, recruited her to pursue a PhD

FIGURE 7.2 Margaret Ellen Mayo Tolbert Credit: Calvin Austin

in chemistry. Her initial support was from the Chemistry Department at Brown University. Later, she applied for and received a fellowship from the Southern Fellowship Fund, which financed her studies at Brown. Teaching night classes in math and science to adults at the Opportunities Industrialization Center in Providence, Rhode Island, provided more funding. At Brown, Professor John N. Fain selected her to work on one of his biochemistry research problems, which was to determine the biochemical reactions that take place in the liver in response to various chemical agents. The energy-bearing materials are stored, processed, and released by the liver in several complicated sequences of chemical reactions. She was able to analyze the reactions and demonstrate how they work. The title of her PhD dissertation is "Studies on the Regulation of Gluconeogenesis in Isolated Rat Hepatic Parenchymal Cells." Mayo's research was so successful that she published three articles in the *Journal of Biological Chemistry* before she graduated.

When she finished her research, she accepted a faculty appointment in the Department of Chemistry at Tuskegee University. Actually, when she enrolled at Brown University in 1970, she received a leave of absence approval from Tuskegee Institute. When she returned to Tuskegee, after completing all of her dissertation requirements at Brown University in three years, she was provided with a letter that indicated that she had completed all requirements for the PhD. This showed that she had the rights and privileges pertinent to any person with the doctorate degree. The official degree was awarded to her in 1974. At Tuskegee, she received research grants so that she could conduct research on campus and supervise that of graduate and undergraduate students, as well as serve as mentor to a selected number of high school students.

She continued teaching and conducting research at Tuskegee until 1977, when she accepted the position of associate professor of pharmaceutical sciences in the College of Pharmacy at Florida A&M University. She became associate dean of the College of Pharmacy shortly after assuming her appointment to the faculty. In 1978, she accepted the opportunity to spend five months in Brussels, Belgium, at the International Institute of Cellular and Molecular Pathology. There she was engaged in research funded by the National Institute of General Medical Sciences that was similar to that which she had completed for her PhD. The research was on how different drugs are metabolized in rat liver cells. When she returned to the United States, she continued her research endeavors for five months in the laboratory of Dr. Elizabeth LeDuc at Brown University as a visiting associate professor. Dr. Tolbert has publications that resulted from her research endeavors.

In 1979, she became the first female director of the Carver Research Foundation of Tuskegee Institute (later University). Tolbert also became the associate provost for Research and Development of the University; these were concurrent positions. Her duties at the Carver Research Foundation included directing research and outreach programs and fund-raising in order to put the foundation on a strong financial basis. She was successful in facilitating campus-wide receipt of research grants and contracts. She also developed a communication strategy that put the foundation into the mainstream of the research community. She built a communications network with institutions of higher learning in West Africa and the U.S. Department of Energy national laboratories (e.g., Pacific Northwest National Laboratory and Oak Ridge National Laboratory).

In 1987, Margaret decided to expand the scope of her managerial experience into industry by taking a sabbatical to work at the Warrensville Heights Research Laboratory of the Standard Oil Company of Ohio. While she was working there, the company, as the result of a merger, became the British Petroleum Corporation of America (BPA) with headquarters in Cleveland, Ohio. She played a key role in the completion of the merger between British Petroleum and Standard Oil. Her role focused on science and technology businesses and communications between the BP Sunbury Laboratory and the BPA Warrensville Heights Laboratory. This role also involved the provision of documents needed to further enable appropriate communications and decision making relative to the merging of the two laboratories. With a title of senior budgets and control analyst and senior planner, she was the highest-ranking African American female at the Warrensville Heights Laboratory of BPA. Even though industry had begun to hire African American scientists after the affirmative action law, promotion to senior-level status took longer than for non-minority scientists. Dr. Tolbert initially went to BP for a year-long sabbatical leave that was facilitated by Dr. Jeanette Grasselli of Standard Oil and was invited by the new vice president for research of BPA to remain at the laboratory once the merger was completed. She remained in this position until 1990.

Ever busy, Dr. Tolbert started to work on the development of science education programs, especially for museums, in her spare time at British Petroleum Corporation. She already had experience in developing this type of program while at Tuskegee. Because of this interest in science education and because of her managerial experience, the National Science Foundation hired her in 1990 as director of the Research Improvement in Minority Institutions (RIMI) Program. She was subsequently appointed to serve as staff associate for Science, Mathematics, Engineering, Technology, and Education (SMETE) and as a member of committees and subcommittees of the Federal Coordinating Council for Science, Engineering, and Technology. Margaret's tenure at the National Science Foundation was three years, the time limit for all rotating program officers. In addition to managing the program, she was the representative of RIMI to Congress. Although she never testified before Congress, she wrote reports to Congress and the public about the program. Margaret was instrumental in ensuring that a larger number of individuals from various minority groups could become award recipients. She also developed and held workshops on the grant proposal process for organizations and educational institutions.

Upon leaving the National Science Foundation, she worked for a short time as a consulting scientist at the Howard Hughes Medical Institute in Chevy Chase, Maryland. There she reviewed research proposals submitted from scientists in the Eastern Bloc countries. Tolbert then accepted a position at Argonne National Laboratories as director of the Division of Educational Programs, becoming the first African American woman to serve in this capacity. There she directed Argonne National Laboratories' educational and training programs that were both national and international in scope. She developed a workshop focused on radiotherapy and cancer prevention, which was implemented in Ghana, West Africa. She also supervised the development of a telecommunication program to train a large number of Illinois teachers in the use of computers, with special emphasis on Internet training. Also, she developed and implemented education and training programs that enabled students, especially minorities, to perform at a higher research level.

In 1996, she became the director of the New Brunswick Laboratory (NBL) of the United States Department of Energy. The New Brunswick Laboratory, which began in New Brunswick, New Jersey, was moved to Illinois between 1975 and 1977.[16] It is an analytical chemistry laboratory primarily focused on measurement technology and the certification of nuclear reference materials and the responsibility for safeguarding nuclear materials. Since its beginning, NBL has maintained its status as a Center of Excellence in the analytical chemistry and measurement science of nuclear materials. In this role, NBL continues to perform state-of-the-art measurements of the elemental and isotopic compositions for a wide range of nuclear materials. Since NBL is a network laboratory of the International Atomic Energy Agency in Vienna, Austria, Dr. Tolbert participated in critical meetings at that agency. She was the first African American and the first female to hold the senior level position of director of the New Brunswick Laboratory.

In September 2002, Dr. Tolbert returned to the National Science Foundation as senior advisor in the Office of Integrative Activities (OIA). OIA supports the Office of the National Science Foundation Director in carrying out a number of cross-organizational activities and provides policy support for these activities. Dr. Tolbert serves as a senior-level agency spokesperson to further promote NSF efforts to increase the participation of underrepresented minority groups, women, and persons with disabilities in science, engineering, research, and education, particularly across all centers and facilities programs for the National Science Foundation. She provides leadership in the oversight of the congressionally mandated Committee on Equal Opportunity in Science and Engineering while acting as its executive secretary and as National Science Foundation's executive liaison to the Committee. As senior advisor, Dr. Tolbert provides leadership and oversight for mission-critical NSF-wide programs.[17]

She has received numerous awards for her contributions to society. She was an honoree for the National Science Foundation's "Behind the Glory is a Story" Observance in celebration of National Women's History Month (2010). In 2007, she was awarded the Dr. George Washington Carver Distinguished Service Award by Tuskegee University. The American Association for the Advancement of Science (AAAS) honored her by appointing her as an AAAS Fellow in 1988. Also included among her awards are the following: Women of Color in Government and Defense Technology Award in Managerial Leadership (July 2001); Letter of Congratulations on Winning the Award in Managerial Leadership (Secretary Spencer Abraham, July 13, 2001); Performance Awards for fiscal years 1999, 2000, and 2001 from the Chicago Operations Office of the U.S. Department of Energy; Performance Award for fiscal year 2005 from the National Science Foundation; Chicago-Tuskegee Alumni Club President's Merit Award (1999); Secretary of Energy Pride Award for Community Service (1998); and Certificate of Distinguished Service to the Federal Reserve System and excellent contributions to the Bank and to the economic progress of the Sixth Federal Reserve District (1987).

Dr. Tolbert is a member of the American Chemical Society, the American Association for the Advancement of Science, New York Academy of Sciences, and the Institute of Nuclear Materials Management.[18] She has completed several tours of duty in foreign countries, including Ghana, Liberia, Libya, Sudan, South Africa, Senegal, Austria, and Belgium.

In addition to having such an impressive career, Dr. Tolbert managed to have an active home life. She was married and divorced twice. Her first husband was Lawson Boila Johnson of Monrovia, Liberia (West Africa). They were married in 1967 and divorced in 1969. She had one son from this marriage, named Lawson, who was born in 1969. Her second husband was Dr. Henry Hudson Tolbert of Chicago, Illinois, whom she married in the chapel on the Brown University campus in 1972. They were divorced in 1978 in Tallahassee, Florida. While they were married, Dr. Henry Hudson Tolbert adopted Lawson and gave him his last name, Tolbert. Margaret Mayo Tolbert managed being a single parent with a very demanding career by sending her son to be raised by the Cook family for a few years when he was young and she was either living abroad or earning her PhD degree. To give Lawson an educational advantage, Margaret sent Lawson to private schools: St. Joseph Catholic School in Tuskegee, Alabama, and St. Paul's School (a boarding school) in Covington, Louisiana. For short periods, he was enrolled at non-private schools in Tallahassee, Florida, and Tuskegee, Alabama. Lawson currently has two sons, Jordan and Jahvon.

About her multi-phased career, she said, "At each stage of my career, I encountered interesting paths, some of which I explored. It is important that educators and policy makers in a broad spectrum of fields communicate as polices are developed. Our society is rapidly changing to an interdisciplinary one, and the amount of information that is available is increasing in leaps and bounds. No discipline should stand alone; education policies have impact on all disciplines. Students of today must be prepared for the jobs and education programs of tomorrow; we must be forward thinking and must prepare them for a world where the line that divided the disciplines in early times are blurring."[19]

CHERYL L. SHAVERS[20]

Dr. Cheryl L. Shavers represents a woman who started life with few choices. After being raised in poverty, she rose to hold numerous high-level engineering and managerial positions at several Fortune 500 companies in the semiconductor industry and later was confirmed by the 106th Congress as the first African American Undersecretary for Commerce for Science and Technology in the Clinton administration.

Cheryl L. Shavers was born in 1953 in Texas. She was raised in Phoenix, Arizona, by her mother, along with her older sister. They lived in a two-room duplex in a predominantly black neighborhood. Shaver's mother never finished high school and worked as a domestic to support her girls. "Her dream for her daughters was that they graduate from high school and grow up to work in air-conditioned offices."[21] Being raised by a single parent meant sewing her own clothes throughout her school years and experiencing hardship and shame as her mother struggled to make the rent.

While in middle school, Cheryl, along with some other neighborhood kids, collected empty soda bottles and gave the money redeemed from the recycler to a street-hardened prostitute in her neighborhood named "Miss Anne." One day Miss Anne was found murdered, and Cheryl became curious about the workings of the criminalists at the crime scene. Desperately wanting a better life and determined not to travel down the same road as Miss Anne, this event sparked a desire in Cheryl to learn more. This later inspired her interest in science and started her on a quest to work for a police crime lab, even if it meant getting a chemistry degree to get there. She was already an excellent student and an avid reader, with a tendency not to simply complete her schoolwork by turning in only a written report but also to demonstrate a desire to experience the subject matter by making a working model of whatever she was writing about. Her curiosity of the world around her became insatiable.

Cheryl graduated from South Mountain High School and received a scholarship to attend Mesa Community College. After receiving her AA degree, she transferred to Arizona State University, where she majored in chemistry. She received an internship to work in the Phoenix Police Department's crime lab, which was her dream job. In the lab, she worked with other criminalists to help develop a method for the separation of blood enzymes from trace materials, later known as enzyme typing. After results from this new method were successfully used in a murder trial, she suffered her first political setback in the workplace.

The lab director, invoking procedural protocols, took her away from conducting supervised lab experiments and assigned her to a series of menial tasks (e.g., washing police cars, picking up dry cleaning, janitorial errands, etc.) just four months before graduating with her BS degree in chemistry. However, unbeknownst to her, this single event would help shape her future in ways she could have never imagined. Humiliated and having lost what she worked so hard to achieve, she quit her job at the crime lab with much consternation.

She later graduated with a BS degree from Arizona State University in 1976 and took a job in the engineering rotational program with Motorola Corporation in Phoenix. As a process engineer, she specialized in discrete device applications. While Cheryl Shavers never envisioned going on to graduate school, engineers in this rotational program are required to spend at least one year in any graduate program of their choosing. She decided to take several business courses in the MBA program at Arizona State University but soon missed the rigors of scientific discovery. Fascinated by a graduate level thermodynamics course that she took "for fun," she did so well that the professor offered her a stipend to pursue a PhD degree. Taking a leave of absence from Motorola and working on her PhD studies at Arizona State University in solid-state chemistry, she received this degree in three and a half years in 1981.

After receiving her PhD degree, Shavers accepted a position at Hewlett-Packard (HP) as a process engineer and moved to the San Francisco Bay area. While there, she painfully learned that her first supervisor managed by traditional methods rather than innovative ones. However, it was his discomfort with "diversity of thought" that

FIGURE 7.3 Cheryl L. Shavers Credit: Photo courtesy of Cheryl Shavers

ultimately hindered her professional progress. As a result, Cheryl moved to the Technical Legal Department at HP and became a patent agent, which required that she attend law school. However, after receiving acceptance offers from three law schools, she soon discovered the monotony of patent drafting. Realizing that it was time to stop doing what she did best—getting degrees—and start taking some risks, she came to a professional crossroads. Four years of law school at night while holding a full-time day job was not going to get her where she wanted to go. She loved science and technology and everything that came with it. At this point, she had to make a choice: stop simply working hard to get the next job or start developing a career strategy in the semiconductor industry by learning as much about the industry as possible. She chose the latter.

Shavers secured a position with Wilton Company as a manufacturing manager of the diode microelectronics lab. This was a risky move because leaving a Fortune 500 corporation to work for a much smaller company could either propel or stall her career. Fortunately, this move afforded her more flexibility to explore her technical and "soft" skills. In this entry-level managerial position, she acquired leadership skills by working alongside her employees and teaching them to explore opportunities through risk taking, as she continued to demonstrate. When Wilton decided to relocate out of the area, rather than go with the company, she decided to take a position with Varian Associates in thin films applications research. Moving back into engineering research enabled her to understand the semiconductor equipment manufacturing sector, where she later earned a promotion to engineering lab director of the Thin Film Development Lab. After Cheryl and her team successfully solved a serious manufacturing problem for Intel Corporation, the largest manufacturer of semiconductors worldwide, she was recruited to join their Components Research team. While at Intel, she not only learned how to strategically manage her career, but she also learned when and how to compete. Risk taking and collaboration with coworkers became her calling card. She earned a reputation for doing the hard work and getting the job done. This led her to numerous promotions and a meteoric rise to high-level leadership positions at Intel. In doing so, she became the highest-ranking black woman at Intel Corp and a prominent female leader in Silicon Valley.

During this time she wrote a weekly column in the *San Jose Mercury-News* entitled "Women In Technology," sharing her own personal five-point work code;

1. No one owes you a living;
2. Be realistic about your work environment and coworkers;
3. Never allow your insecurities to influence your actions (i.e., create a plan to win);
4. Value people for their accomplishments;
5. Take risks.[22]

In 1999, she was nominated by President Bill Clinton to become Undersecretary of Commerce for Technology at the U.S. Department of Commerce where she oversaw the Commerce Department's Technology Administration (TA), the Office of Technology Policy (OTP), and also the National Institute of Standards and Technology (NIST). In this government position, Dr. Shavers served as senior advisor to the Secretary of Commerce in forming new policies and program initiatives in the areas of science and technology. She also represented the administration on the National Science and Technology Council's Committee on National Security, Committee on International Science, Engineering and Technology, and Committee on Technology. Internationally, Dr. Shavers served as a member of the U.S.-Israel Science and Technology Commission (USISTC), served as a co-chair of the USISTC Joint High Level Advisory Panel, served on the Board of Directors of the U.S.-Israel Science and Technology Foundation, was former co-chair of the Technology Subcommittee under the U.S.-Egypt Partnership for Economic

Development, and was a frequently invited participant on the U.S.–South Africa Science and Technology Committee, U.S.-Russia Science and Technology Committee, U.S.-China Joint Management Committee, and U.S.-Japan Joint High Level Committee.

Dr. Shavers is currently the CEO of Global Smarts, Inc., a business services firm that specializes in advisory and business accelerator services for asset value extraction and leadership exchanges. She is a member of the board of directors of Rockwell Collins Company, a leader in the design, production, and support of communications and aviation electronics solutions for government and commercial customers worldwide, as well as a member of the board of ATMI, a leading supplier of materials and packaging products to the semiconductor industry. She is also an adjunct professor at the Leavey School of Business at Santa Clara University, a private Jesuit institution, and a Henry Crown Fellow at the Aspen Institute.

Formerly, she was a board member of Intel Foundation's "Women in Science and Engineering Scholarship" program, a member of the board of directors of the San Jose Tech Museum of Innovation, a mentor for the Semiconductor Research Corporation, and was the producer and host of *Technology's Edge*, a radio talk program broadcast on Business Radio, AM570—WWRC in Washington, D.C. She also served as a board member of the Anita Borg Institute, a nonprofit dedicated to expanding opportunities for women, girls, and minorities in science and technology.

Dr. Shavers has won numerous prizes and awards. In 2010 she was an honoree for promoting diversity in the boardroom by Boardroom Bound and she was featured in "Directors to Watch in 2009" by Directors and Boards

Advisory Annual Report 2009. Additionally, she received the 2005 Compass Award from the Women's Leadership Exchange and received Congressional recognition from the U.S. House of Representatives in 1999 and 2005 for her work in the community and for her business leadership.

In 1996 she was one of ten inductees into the inaugural class of the International Women in Technology Hall of Fame and was later inducted into Arizona State University's College of Liberal Arts and Sciences Hall of Fame. Dr. Shavers lives in the San Francisco Bay area with her husband, Joseph C. Agu, daughter, and twin sons.

CHAPTER 8

Chemical Engineers

LILIA ANN ABRON

Dr. Lilia Abron is an engineer, an entrepreneur, mother, and activist who works twelve-hour days. She is another true Renaissance woman.

Lilia was born at home in Memphis, Tennessee, on March 8, 1945. She was small, premature, and almost did not survive were it not for her aunt, who rushed her to the hospital in a cab because ambulances were not available to black people at the time. She was the second of four daughters of Ernest Buford Abron and Bernice Wise Abron, who were both educators. Both of her parents had attended LeMyone College. Her father entered college and played football. Because of an injury he was ineligible to serve in the military in World War II. He then worked as a Pullman porter, because his father had been a Pullman porter. After the war, when the trains were not as popular, he became a teacher in the Memphis public schools. Lilia's mother and father were very active during the civil rights era. Lilia's mother was from Arkansas; and she typed the briefs for Wiley Branton, defense attorney for the Little Rock Nine, the group that integrated Central High School in Little Rock, Arkansas.[1]

Because Lilia's parents were active in Memphis society, Lilia was involved in programs that included the Girl Scouts and the church. She went to public school in Memphis, Tennessee. In 1957, the Soviet Union launched Sputnik, which led the United States to improve math and science education. The school system tracked each student's education, even in the segregated schools. Therefore, Lilia was placed in the math and science track. This meant she participated in a science fair, which was held at Lemoyne College. In addition, she had to prepare other science projects. Her segregated schools were well equipped for science teaching. In addition to well-stocked labs, the Memphis high school that she attended offered higher-level mathematics, including algebra and introduction to calculus. She graduated from high school in Memphis and decided to go to college with the intention of studying medicine, which was the one of the few occupations available to black people at the time. Lilia's oldest sister majored in biology at a college out of town. Since her parents could not afford to send two students to an out-of-town college, Lilia's only choice was to stay at home and attend in town at LeMoyne College. The historically black, small liberal arts college had no dormitories, so the students commuted to college and the out-of-town students stayed with families in town. She liked the atmosphere of LeMoyne. There were about only 500 students in the college. Everyone knew each other because most of the students were from Memphis. She later said that colleges like LeMoyne have a mission to serve students by helping them to mature so that they can go on to graduate school.

She entered LeMoyne with a scholarship in her freshman year, which she lost because she did not keep up her grades (but her parents paid the tuition). She majored in chemistry because in freshman biology she did not like working with the animals. Lilia's advisor at Le Moyne, Dr. Beuler, persuaded her to study engineering. Because of this decision, she wanted to switch to another college in town, which had a greater emphasis on engineering. Despite majoring in chemistry, she did not like working in a lab. She discovered chemical engineering through reading magazines about the career and visiting some of the white colleges in the area. She also took all the liberal arts courses in reading and had almost a math major as well. The decision to study engineering seemed somewhat risky; at the time, the people whom she knew who had degrees in engineering were not pursuing their career but were working in the post office. Because of this, her parents wanted her to take some education courses as a backup, but she did not want to teach. She wanted to leave Memphis, and she felt that engineering was a way out of Memphis. She graduated from LeMoyne in 1966 with a BS in chemistry with distinction (a designation that indicated that the faculty felt she was a person who helped other students).

Since she did not take the education courses that her parents wanted her to take, she knew she had to go to graduate school. She looked at the posters in the college about engineering and found one about sanitary engineering; she decided to apply. At the time, engineering students coming from historically black colleges were being advised to go to a fifth-year program; someone at her college told her that it meant that major colleges did not think that these students had received a good education. She was advised to go to a regular graduate school instead of a college of engineering. She applied for the Peace Corps and to other graduate schools for graduate studies. One of them, Washington University in St. Louis, offered her a scholarship in sanitary engineering. A full-tuition scholarship that was funded by the Public Health Service, a part of the Department of Interior, for environmental engineering was a breakthrough. She had never had a full-tuition scholarship before.

In the fall of 1966, Abron entered Washington University, which was a very different world. Away from home for the first time, she found the atmosphere different, with larger class size, less personal attention, and classmates who were predominately white males. There was only one other woman in the program. Her white male colleagues did not believe that she was there to get the degree and become a professional engineer. They thought that she would just get married and have babies. She proved that she could do it by completing the work for the degree in thirteen months. She said it was because she had nothing else to do but work on her degree. The experience opened the world of academia to her, and she developed a new goal: to work in academia, eventually reaching the level of college president. She worked in the labs of her professors while in graduate school. She graduated from Washington University with an MS in environmental engineering in 1968.

Even with an MS degree in environmental engineering, she found it hard to get a job as an environmental engineer. Companies were not ready to take on a woman engineer, never mind a black women engineer. Companies would say they did not even have a ladies' room! Therefore, from 1967 to 1968 she worked as an environmental chemist at the Kansas City, Missouri, water department. This was laboratory work, which she did not like. She changed jobs in 1968 to work as a research environmental chemist at the

Metropolitan Sanitary District of Chicago, where she stayed for about a year.

In September 1968, Abron entered the University of Massachusetts in the Department of Civil Engineering to work with Dr. Rolfe Skrimbee to pursue her PhD. Her advisor then moved from U. Mass to the University of Iowa and she went with him. She found the atmosphere at the University of Iowa welcoming. Iowa City, home of the university, is a college town. Everyone tried to make her stay as comfortable as possible. Later, her advisor left the University of Iowa while she was still working on her research in sanitary engineering. The Sanitary Engineering Department was taken over by the Department of Chemical Engineering. Since she was too far along in her research and education process to move, she transferred to the Department of Chemical Engineering and had to take a few more courses to obtain her degree in chemical engineering. Her research involved reverse osmosis, and the equipment was housed in the labs of the Department of Chemical Engineering. The title of her thesis is "Transport Mechanism in Hollow Nylon Fiber Reverse Osmosis Membranes for the removal of DDT and Aldrin from Water." She received her PhD in chemical engineering in 1972.[2] She is the third woman to receive a doctorate in chemical engineering from the University of Iowa.

Since her goal now was to achieve in academia, Dr. Abron's first appointment in 1971 was as an assistant professor of civil engineering at Tennessee State University. In 1973, she held a joint appointment as an assistant professor of environmental engineering at Vanderbilt University. She held both positions until 1975 when she left for a position as assistant professor of civil and environmental engineering at Howard University. While at Howard, she also held a joint position at Washington Technical Institute (now a part of the University of the District of Columbia). While at Howard, she learned that many of the faculty members in her department owned engineering firms. Her colleagues could not envision an engineering firm owned by a woman.

Undaunted, in 1978 she founded PEER Consultants PC, an environmental and sanitary engineering consulting firm. In the 1980s her company struggled, as did other companies during the recession. She had to learn what it took to keep her company alive during this recession. PEER was awarded contracts from the Superfund program, which included the Boston Harbor cleanup; the Department of Defense for environmental policy work; and the Department of Energy through its Hazardous Waste Remedial Actions Program. *Engineering News-Record* magazine ranked PEER as a top engineering and design firm, for its longevity remarkable in a field that saw many minority firms fail. In 1996, PEER received the Department of Defense Small Business/Environmental Restoration Excellence Award for work done over many years for the U.S. Air Force and the Air National Guard. The company is headquartered in Rockville, Maryland; PEER has offices in seven U.S. cities and has had a staff of one hundred workers.

In 1995, Abron founded PEER Africa Pty. (Ltd.) with co-owner Douglas Guy in Johannesburg, South Africa. After the official end of apartheid in South Africa in 1994. houses were needed for millions of people. PEER Africa designed energy-efficient homes—40 to 50 percent more efficient than traditional homes—that could be built for less than $10,000. PEER Africa worked on projects elsewhere in Africa, including Mali, Uganda, and Nigeria, as well as the Caribbean islands. In 1997, the United Nations Framework Convention on Climate Change praised PEER Africa's trademarked methodology, known as the PEER Africa Energy and Environmentally Cost Optimized Human Settlement Development Model. The United Nations presented Abron with an award for creating the innovative, environmentally friendly design for these South African homes.

The housing program developed by PEER Africa was designed to create holistic, economically viable, sustainable, and environmentally friendly communities. The initial target was the majority (black) population, even though all of South Africa would benefit. The prototype developed for this project is called the ECOTM house. It incorporates passive solar design, energy efficient appliances, and environmentally friendly technologies such as passive solar water heating and water conserving devices. The goals of this project were to promote capacity building at all levels, to create jobs, and to develop and support businesses in the construction and building supply industry and associated markets. Another goal was to develop technical managerial and administrative leadership within the townships.

Dr. Abron is a registered professional engineer and a member of several professional and technical societies, such as the Water Environmental Federation, American Society of Civil Engineers, American Water Works Association, Society of Sigma Xi, and American Association of University Women. She serves on the Engineering Advisory Board for the National Science Foundation.

Dr. Abron has received numerous awards, among them: Magic Hands Award by Le Moyne-Owen College

(May 2001); American Academy of Arts and Science (2004); Alumni Achievement Award, Washington University School of Engineering and Applied Science (2001); induction into the National Black College Alumni Hall of Fame (1999); Hancher Medallion, presented by the Finkbine Committee of the University of Iowa for learning, leadership, and loyalty to the university (1999); admission to the Engineering Distinguished Alumni Academy at the University of Iowa (1996); and William W. Grimes Award for Excellence in Chemical Engineering from the American Institute of Chemical Engineers (1993).

In addition to her professional work, Dr. Abron is very active in the community. She uses her business acumen and engineering expertise to promote science education in primary and secondary schools. Her company supports science fair projects by purchasing equipment and supplies for students. She encourages PEER staff to work with children in their neighborhood schools, and she serves as a mentor for several students each year. She is also active on the lecture circuit, visiting colleges to serve as a role model and mentor to students.

She is active in her church, playing handbells in the Angelus church choir. She is a board member of the Baptist Home for Children and has served as president of the Washington, D.C., chapter of Jack and Jill of America.[3] In addition to graduate study and work, Dr. Abron took time to become a wife and mother. She was married to Robinson; they have three sons, Frederick, Ernest, and David. She later divorced her husband. Her three sons all attended college, and she guided them through the dangers inherent in being young African American males in the community. She is proud of that.[4]

JENNIE PATRICK

Jennie Patrick is the first African American woman to earn a PhD in traditional chemical engineering. She has also been active in the African American fight for racial and environmental justice.

Born on January 1, 1949, in Gadsden, Alabama, Patrick was the fourth of five children born to James and Elizabeth Patrick. Her parents had no education beyond the sixth grade. Her father worked as a laborer, and her mother worked as a housekeeper. Patrick later recalled, "We lived a very simple life. During my childhood, my father maintained a small chicken farm and a few large vegetable gardens, in addition to working as a laborer during his working years. The fact that my parents had very little formal education did not hinder them from encouraging all their children to pursue a formal education."[5]

According to Patrick, her parents "were disciplined and willing to make necessary sacrifices to assure a better life for all their children."[6] Believing that education was the way out of poverty, Patrick's parents saved up their meager earnings to purchase two sets of encyclopedias for their children, which Jennie frequently read. She frequented the local library to discover other kinds of books. Patrick enjoyed a nurturing educational environment during her early years. She had the good fortune of having several African American teachers who made lasting impressions on her by challenging her to search for knowledge, using her own abilities to achieve whatever she wanted. Growing up in the rural South, her elementary and middle schools were segregated. When she was of high school age, in 1964, she integrated Gadsden High School, as the *Brown v. Board of Education* ruling continued to be implemented in the South.[7] Jennie decided to attend the previously all-white Gadsden High School because it had all the scientific equipment she needed for her studies, while the school for African Americans had very little. She says, "The initial months were a living nightmare. The emotional, psychological, mental, and physical violence against us was difficult to comprehend."[8] She suffered from discrimination by not only the students but the teachers as well. She did not tolerate their abuse and met force with force. She had to challenge the teachers and counselors to be placed in college preparatory courses in high school. In spite of all the problems, she graduated with honors in 1967.

Upon graduation from high school, Patrick received a scholarship to attend the University of California at Berkeley. However, her mother did not want her to go so far away. Therefore, she attended Tuskegee Institute (University) in Alabama. She knew she wanted to major in a technical field, but was not sure which one. She decided to major in chemistry in her freshman year. At the end of her freshman year, Tuskegee started a major in chemical engineering, so she changed majors. As a child, she had known that engineers were the people who designed and made things, and this intrigued her. For reasons unknown to her, the chemical engineering program at Tuskegee collapsed after two years. She decided to leave Tuskegee and go to Berkeley. The scholarship at Berkeley was not available to transfer students, so she had to work for a year to earn the money for tuition. When she entered Berkeley, she had barely enough money to survive. She worked as an assistant engineer at Dow during the summer of her junior year of study at Berkeley. The following summer she worked at

Stauffer Chemical Company in order to support her financial needs. She suffered extreme discrimination at Berkeley from many faculty and students who verbalized their belief that she did not belong there. Patrick expressed that she did not receive much encouragement at Berkeley. She believed the focus of many was instead to discourage and prevent her from being successful as a student. She graduated from Berkeley with a BS in chemical engineering in 1973.

Patrick went on to MIT to pursue a PhD in chemical engineering. She chose MIT because of its reputation as being one of the best engineering schools in the country. MIT proved to be a better environment for black students interested in science and engineering. The number of blacks

FIGURE 8.1 Jennie Patrick Credit: Photo courtesy of Jennie Patrick

at MIT was relatively small but was nevertheless greater than at other institutions at the time. She sincerely enjoyed the intellectual challenge at MIT, but she had to overcome racial prejudice. The subjects on which she concentrated were thermodynamics, homogeneous nucleation, heat and mass transfer. She was a research assistant as well. She graduated from MIT with a PhD in chemical engineering in 1979. She is the first African American woman to earn a PhD in "traditional" chemical engineering. Dr. Patrick is featured in many children's books. She has also influenced many women of various backgrounds to pursue chemical engineering careers.

Her first position after graduate school was at General Electric Research in Schenectady, New York. While at General Electric she focused on research and development projects. She developed a pioneering research program in the area of supercritical fluid technology. Supercritical fluid extraction technology is a high-pressure separation and purification technology, which combines the unit operations of distillation and liquid-liquid extraction. Her expertise in supercritical fluid technology provided Patrick with numerous career opportunities. After three and a half years with General Electric, she joined Philip Morris Research Center in Richmond, Virginia. She headed their newly established supercritical extraction technology program. She also designed a state-of-the-art supercritical extraction pilot plant facility.

As she continued to advance her career, in 1985 she moved to Rohm and Haas Company Research Laboratories in Bristol, Pennsylvania, where she became a research section manager at Rohm and Haas. As a manager, Dr. Patrick was involved in other technical areas such as polymer science and emulsion technology, as well as supercritical fluid extraction. In time she became the research manager to the company's first fundamental research engineering group. In this position, she was involved in not only the technical aspects but also the administrative and managerial responsibilities.

In 1990, after five years at Rohm and Haas, she became assistant to the executive vice president at Southern Company Services in Birmingham, Alabama. She held this managerial position until 1993, when she decided to pursue a career in academia. Patrick had previously been an adjunct professor at two universities, Rensselear Polytechnic Institute (1980–1983) and Georgia Institute of Technology (1983–1987). It was natural for her to return home to Tuskegee Institute (University) in 1993 to the restored Department of Chemical Engineering as the 3M Eminent Scholar/Professor. The 3M endowed chair was sponsored by the state of Alabama and the 3M Company of Minneapolis, Minnesota. In addition to her research projects at the university, she was devoted to helping minority students succeed in science and engineering careers. She said, "Being here (at Tuskegee) provides me the opportunity to make a difference in the lives of students. I've been able to teach them and share many of my experiences with them—thus providing them hopefully with some insight into what the future may hold for them."[9] She held this position until 1997, when she became a senior consultant at Raytheon Engineers and Constructors in Birmingham, Alabama, where she stayed until the year 2000.

Currently, she is the consultant for Education & Environmental Solutions in Peachtree City, Georgia, a firm that she founded. In 1980, she received the National

Organization for the Professional Advancement of Black Chemists and Chemical Engineers (NOBCChE) Outstanding Women in Science and Engineering Award. The CIBA-GEIGY Corporation in 1983 featured her in its Exceptional Black Scientists Poster series. Patrick is a member of the American Institute of Chemical Engineers, Sigma Xi, and NOBCChE.

Dr. Patrick is married to a physician who is a former practicing chemical engineer. They have no children. Her hobbies include gardening, which she calls her quiet time, where she communes with nature. Her advice to young people contemplating a career in science is, "You need to know who you are, you need to be comfortable with yourself, you need to love yourself, and you need to respect yourself. Then everything else becomes secondary. Achieve the highest goals possible but don't allow achievements alone to define who you are. Make the decision early. Don't let material things or world recognition be your driving force."[10]

CHAPTER 9

My Story

JEANNETTE ELIZABETH BROWN

Jeannette Brown's career has included accomplishments in industry, academia, and publishing. Her claim to fame is working in two different pharmaceutical firms, where she was able to contribute her skill to the research teams who produced several marketable drugs.[1] She was also able to mentor minorities to encourage them to enter the field of chemistry, both as part of a corporate effort and as a volunteer.

Jeannette Brown was born May 13, 1934, in Fordham Hospital in the Bronx, New York. She was the only child of Ada May Fox and Freddie Brown. She was born in the middle of the Depression, and times were tough. Her father worked a number of jobs in order to feed his family, including shining shoes on the street. Finally, when Jeannette was five, her father got a job as a superintendent in a building in the Washington Heights section of Manhattan. This section of Manhattan was just becoming a home for middle-class blacks moving up from Harlem. Since her father was a super, he had a basement apartment in the building. One of the tenants in the house was Dr. Arthur Logan,[2] who became Jeannette's doctor when she became very ill.

Jeannette was in and out of the hospital many times, and she remembers asking Dr. Logan how she could become a doctor. He told her that she would have to study science. Jeannette was only five or six at the time, but that conversation impressed her and she immediately decided to become a scientist. When Jeannette started school at the age of six, she went to the neighborhood public school, which all children did at the time. The children in the school were mostly black, and some of them taunted her because she was interested in being a good student. Her father decided that the only way that she was going to get a good education was for him to try to get a job as a superintendent in a white neighborhood so that Jeannette could go to the mostly white schools. This is what he did. The family moved from Manhattan to the Bronx and finally Brooklyn as his jobs changed. Soon he had two jobs: he was employed by the postal service, but he continued to work as a super for the housing.

Jeannette was a good student, and when she entered junior high school, she was placed in a "rapid advance" class so that she could skip a grade. Her mother decided that that would not be a good idea for her because of her language arts skills; she was good at science and math but not English. So she was put into a regular class. When she was in her junior year in high school, the family moved to an apartment in a housing project in Staten Island. Her father no longer had to work two jobs. She took chemistry in New Dorp High School and was challenged by the teacher to get a grade of 100 percent on the New York State Regents exam.[3] When Jeannette graduated from New Dorp High School in 1952, her name was placed on the permanent honor roll of the school. She applied to three colleges: Wheaton College in Illinois, Barnard College in New York, and Hunter College in New York. Although she was admitted to the colleges, none gave her the scholarship money that she needed to attend. So she went to Hunter College in New York, one of the City of New York colleges (now City University), and at that time, the only all-female college of the New York City College system. The tuition was free, as were all the City Colleges. Because she declared a major in chemistry, she was put into a new chemistry major, called Chemistry Field, which meant that she started her major courses as a freshman. When she took organic chemistry, she met Professor Arthur Sweeny, who later became her mentor. He followed her career for the rest of his life. She graduated from Hunter College in 1956 and decided to go to graduate school to obtain a master's degree. Still focused on medicine, she did not have the money to enter medical school at the time. There were few openings in medical school for non-white students; there were reputed to be only two slots in most medical schools, and students who were veterans or Phi Beta Kappa were preferred.[4] Therefore, Brown decided that she would go to graduate school to obtain an MS degree in order to get a job in industry as a chemist.[5] She applied to two graduate schools for teaching assistantships and was accepted at both; she chose to go to the University of Minnesota for graduate studies. The only black student in her class, she chose as a thesis advisor Dr. Fred Keolch, who took her under his wing. She worked as a teaching assistant for Dr. Wayland Noland, who encouraged her to pursue a PhD degree when she had finished her MS thesis in June 1958. The title of her thesis was "A Study of Ylide Formation in Quaternary Ammonium Salts of 9-p Amino Fluorene."

She graduated in June 1958, becoming the first African American woman to receive a degree in chemistry from the University of Minnesota.[6]

Jeannette's goal was to work in the pharmaceutical industry for five years and then go to medical school. She sent applications to all the pharmaceutical corporations that were listed in the American Chemical Society career list. CIBA Pharmaceuticals (now Novartis) and Schering Plough (now Merck) accepted her. She chose CIBA, where she became the third black scientist to work at CIBA.[7] All of the black chemists who worked at CIBA at the time had MS degrees. While at CIBA, she was able to contribute information for two publications. She worked at CIBA for eleven years and then moved on to Merck because there were more opportunities for advancement.

FIGURE 9.1 Jeannette Elizabeth Brown Credit: Photo courtesy of Merck Research Labs, Merck & Co. Inc.

While at Merck, she worked on several research teams.[8] She was able to work on teams that produced drugs that have been marketed, and she has publications about this research. She is most proud of her work on the team that synthesized the cilastatin sodium component of Primaxin, an antibiotic. The cilastatin sodium is an inhibitor of the kidney enzyme renal dipeptidase, dehydropeptidase, that was inhibiting the potency of the antibiotic Imipenem. The combination of the two components, cilastatin sodium and Imipenem, constitutes Primaxine, which is a potent broad-spectrum antibiotic that has been on the market since 2006.

Brown took an early retirement from Merck and her research career to work in academia as an administrator of the New Jersey Statewide Systemic Initiative (NJSSI) program at the New Jersey Institute of Technology (NJIT). NJSSI was a National Science Foundation grant designed to improve the teaching and learning of science and math statewide. In order to learn more about teaching and learning in the K-12 education system, she applied to West Ed to enter the National Academy for Science and Mathematics Leadership. After a year and a half of study, she became a fellow of Cohort III of the National Academy for Science and Mathematics Leadership. This program gave her the tools needed to conduct teacher education workshops in science. Brown was also instrumental in obtaining a grant from the Camille and Henry Dreyfus Foundation for teacher education workshops in chemistry for middle and high school teachers. In this grant, she hired experienced teachers to teach the summer workshops. Teachers who attended were paid a small stipend, providing they could demonstrate that they were using the information from the workshop. As all the teachers in the program were able to demonstrate the use of the materials, Brown has been able to advance the teaching of chemistry in middle school and high school; she learned a lot about working with the educational system, which deepened her appreciation of the job that teachers do.

In 2002 Miss Brown retired from NJIT and she was able to pursue full-time the study of the history of African American women chemists, begun as a hobby while she was working, In order to do this, she applied for and received two fellowships from the Chemical Heritage Foundation, the Société de Chimie Industrielle (American Section) Fellowship (2004) and an Ullyott Fellowship (2009). This enabled her to do additional research for this book.

Miss Brown has received numerous awards, including being named Fellow of the American Chemical Society (2009), Fellow of the Association for Women in Science (2007), Volunteer of the Year by the Somerset County (New Jersey) Women's Group (2010), and Outstanding Mentor by the Metro Women Chemists Committee of the North Jersey Section of the American Chemical Society (2010).

In 2005 she won the ACS Award for encouraging disadvantaged students into careers in the chemical sciences, sponsored by the Camille and Henry Dreyfus. With the funds from this award, she established the Freddie and Ada Brown Award for African American and Native American middle school and high school students in New Jersey. This award is among the few awards for middle school science students. In order for students to win the award, they must write an essay about why they like science (middle school) or chemistry (high school). The essays are judged by teachers, and the winners, two in each age group, receive a savings bond, a gift certificate, and a book. They also receive a mentor. Miss Brown tries to keep track of the students.

Miss Brown is a member of the National Organization for the Professional Advancement of Black Chemists and Chemical Engineers (NOBCChE); Iota Sigma Pi, a chemistry honor sorority; and the American Chemical Society (ACS). She is very active in the ACS, both locally and nationally. Locally she serves as the publicity director, and nationally she has served on numerous committees and has been elected by the governance to serve as councillor. Miss Brown never married, electing to devote her life to her career and mentoring students.[9]

CHAPTER **10**

Next Steps

This book covers the stories of African American women chemists who entered the field before the civil rights era or shortly thereafter. The women in this book were swimming against the tide. They chose to enter a career in which they were among the first persons of their gender and race. There are many young American women currently active in the profession of chemistry. Young girls are still interested in becoming scientists; however, they still have to fight to make sure they receive a good education in science. This need for good education is discussed in a book titled, *Swimming against the Tide: African American Girls and Science Education,* by Dr. Sandra L Hanson.[1] Dr. Hanson studied young African American girls in high school and their attitudes toward science, which has traditionally been a male profession. One of Dr. Hanson's conclusions is that these young girls had never seen a role model, African American women chemists, either in books or in person. She also discovered that the young women were still interested in science, even though they received little encouragement from their teachers. Dr. Hanson found that the parents of the girls encouraged their interest in science, with mothers being their biggest cheerleaders. Dr. Hanson would like to see more science teachers encourage young African American girls to study science. One of the women in this book, Allene Johnson, served as a high school chemistry teacher and made a career both of mentoring science teachers and of encouraging African Americans, both male and female, to enter careers in chemistry. Most of the other women in this book also have mentored women who wanted to become chemists either in academia or in industry.

What resources are available to young people if they are interested in chemistry as a career? There are many organizations that are focused on careers in chemistry. The major one is the American Chemical Society (ACS). However, one should first consider this question: What is chemistry? Chemistry is a varied profession. It is not just working in a research laboratory making new chemicals. According to the ACS, "In simplest terms, chemistry is the science of matter. Anything that can be touched, tasted, smelled, seen or felt is made of chemicals."[2] As can be discovered by reading about the women in this book, chemists have a variety of careers. The ACS web site states, "Chemists are the people who transform the everyday materials around us into amazing things. Some chemists work on cures for cancer while others monitor the ozone protecting us from the sun. Still others discover new materials to make our homes warmer in the winter, or new textiles to be used in the latest fashions. The knowledge gained through the study of chemistry opens many career pathways."[3] The American Chemical Society web site (www.acs.org/careers) lists thirty different types of chemical careers, and there may be more. Currently these careers are open to anyone who has the qualifications. The American Chemical Society also has information about chemistry for children from kindergarten age to graduate school. Allene Johnson recommends that students learn that chemistry is not hard and need not be feared. Therefore, trying to interest students in chemistry at a young age makes sense. For young students, their parents, and educators, there are several books that contain science activities that can be done by anyone. They are listed at the ACS web site in the education tab "Middle School and Elementary School" and are further categorized by grade. So the seeker will find "Science for Kids," grades 2–6, "Inquiry in Action—Science Teaching Guide (grades 3–6)," and Science Activity Books *Best of Wonder Science*. In addition, there are two books for pre-school students, *Apples, Bubbles, and Crystals* and *Sunlight, Skyscrapers, and Soda Pop*.

For middle school students seriously thinking about a career in chemistry, there is a middle school activities web site, http://www.middleschoolchemistry.com/.

The ACS education tab has a number of resources for high school students and teachers, including *Chemistry in the Community* (*ChemCom*, a textbook focused on teaching chemistry by showing what chemists do). There is *ChemMatters* magazine, which has featured an article about one of the women in this book, Alice Ball. This magazine also has teacher resources.

For high school students, there are chemistry clubs that can be started in their schools as well as a program called Project SEED. Project SEED is a summer research program for economically disadvantaged students. If student is a junior or senior year in high school, he or she can learn how to can work alongside scientists in a research laboratory.[4] Students who are not economically disadvantaged can find summer enrichment programs or internships run by local colleges and universities. Students should check with their local college or junior college to see if there are summer programs for high school students. But they must apply early because registration is sometimes limited.

The ACS also has programs for students when they enter college. Since funding for college might be a problem, the ACS started the ACS Scholars Program, which provides renewable scholarships for underrepresented minority students who want to enter fields in the chemical sciences. Scholars are selected based on academic standing, financial need, career objective, leadership skills, and involvement in school activities and community service.[5] Information about other scholarships is available on the ACS web site under the funding and awards tab.

To be eligible for a Project SEED scholarship, high school senior students must have worked at least one summer at a science institute under the Project SEED program. Scholarships are restricted to students who plan to major in a chemical science or engineering field such as chemistry, chemical engineering, biochemistry, materials science, or some other closely related major. The scholarships are intended to assist former SEED participants in their transition from high school to college. Project SEED scholarships are nonrenewable and are only awarded to first-year college students. Selection is based on achievement in school, success in the Project SEED program, financial need, and intended chemical-related field of scientific study.

There are scholarships for prospective teachers of chemistry. The ACS-Hach Second Career Teacher Scholarship is awarded to professionals in the chemistry field pursuing a master's degree in education or becoming certified as a chemistry/science teacher. This is intended to improve the teaching and learning of chemistry in high school.

For pre-service teachers there is the ACS-Hach Land Grant Undergraduate Scholarship, awarded to undergraduate chemistry majors who attend a land grant college and express an interest in teaching high school chemistry. Students are selected on the bases of chemistry aptitude, interest, and need.

Even after students graduate to become chemists or chemical engineers, the ACS has many resources for them. There is a Women Chemist Committee (WCC) that focuses on all women in chemistry. The Women Chemist Committee web site, http://womenchemists.sites.acs.org/, contains information and links to other organizations. One of those links is: *Women-Related Web Sites in Science/Technology.* This link goes to many other web sites useful for young girls interested in science. Currently the WCC is focusing on women of color in chemistry, but more about that later. The ACS also has a Committee on Minority Affairs (CMA) for minority chemists. It works to attract minorities to the profession and to make sure that the chemistry community recognizes them for their accomplishments. Information about CMA is on the ACS web site under the governance/committees tab. Last but not least, the Younger Chemist Committee (YCC) focuses on the needs of all chemists under the age of thirty.

One organization focuses specifically on the needs of African American chemists—the National Organization for the Professional Advancement of Black Chemists and Chemical Engineers (NOBCChE).[6] An Ad Hoc Committee for the Professional Advancement of Black Chemists and Chemical Engineers was organized in April 1972 to establish an organization that would focus on the needs and careers of black chemists. That organization's first annual meeting was held in 1973 and it has been going strong ever since. The organization meets every year and awards grants to African American students. They have programs for middle school and high school students. There is a list-serve of NOBCChE for the women members of the organization. Recently they have reconnected with the American Chemical Society for joint programs. According to their site, "NOBCChE is committed to providing opportunities to broadening the pipeline to STEM (Science, Technology, Engineering and Mathematics) careers and opportunities. Each year during the National Conference the organization sponsors a Science Fair and Science Bowl." The Science Fair provides a venue for the students to present original research through a poster competition in which students compete individually. The Science Bowl allows students to compete in teams of four against other teams in a round robin academic quiz bowl. What makes this competition unique is that at least 20 percent of the questions are about African American inventors, scientists and engineers. Both competitions are divided into Junior (6th–8th grade) and Senior (9th–12th grade) divisions. Trophies, along with other prizes, are awarded to students in each level for both competitions. Students are honored at an awards luncheon, along with a science educational enrichment activity.[7]

The U.S. Department of Education sponsors Upward Bound grants that colleges can apply for and run for students. One of the women in this book, Lynda Jordan, is a product of one of those programs. Upward Bound provides fundamental support to participants in their preparation for college entrance. The program provides opportunities for participants to succeed in their precollege performance and ultimately in their higher education pursuits.

Upward Bound serves high school students from low-income families and high school students from families in which neither parent holds a bachelor's degree. The goal of Upward Bound is to increase the rate at which participants complete secondary education and enroll in and graduate from institutions of postsecondary education. All Upward Bound projects must provide instruction in math, laboratory science, composition, literature, and foreign language.[8]

At least one African American Sorority has a program for African American girls. It is Delta Sigma Theta (http://www.deltasigmatheta.org/index.htm). This group has a National Science Foundation to form the Delta Academy for Math and Science: "Catching the Dreams of Tomorrow, Preparing Young Women for the 21st Century." The Delta Academy was created out of an urgent sense that bold action was needed to save our young females (ages 11–14) from the perils of academic failure, low self-esteem, and crippled futures. Delta Academy provides an opportunity for local Delta chapters to enrich and enhance the education that young teens[9] receive in public schools across the nation. Specifically, they augment their scholarship in math, science, and technology, as well as their opportunities to provide service in the form of leadership through service learning, and their sisterhood, defined as the cultivation and maintenance of relationships. A primary goal of the program is to prepare young girls for full participation as leaders in the twenty-first century.

The Alpha Kappa Alpha sorority (http://www.aka1908.com/) also has a program for young girls: Emerging Young Leaders (EYL) Initiatives. This signature program will impact the lives of 10,000 girls in grades six through eight by providing leadership development, civic engagement, enhanced academic preparation, and character building. The increasing demands of the twenty-first century mandate our youth to be better leaders at a younger age, making smart choices with positive consequences.[10]

Iota Sigma Pi is a national honor society for women in chemistry. Its major objectives are: to promote interest in chemistry among women students; to foster mutual advancement in academic, business, and social life; and to stimulate personal accomplishment in chemical fields. Iota Sigma Pi was founded in 1902 and was organized on a nationwide basis in 1916. They sponsor awards for chemists. The number of African American women in this organization is not known, but there are a number who are active members.

If one becomes an academic chemist, there is an organization that was founded to help women in academia. It is called COACh and was formed in 1998 by a group of senior women faculty in the chemical sciences from across the United States with a common concern about the gender-based obstacles that women scientists face in trying to attain their career goals. They sponsor workshops to help women get ahead in their careers.[11]

For historians of either African Americans or of science or of women, or any combination of the above, there are resources as well. Many colleges have courses for those subjects. I have seen curriculum lists for courses from several colleges that are teaching the history of African Americans in science. However, most of them list only men. (In fact, I was contacted by an African American woman who was taking such a course, asking me for a woman whom she should research.) The same goes for the history of women in science. Most published sources neglect the women in this book. I have been asked by college professors to write this book so that they can use it for their course. Some public school districts also teach the history of African Americans in science. Some states, New Jersey being one, are writing curricula for this to be taught in the schools. I have listed some of the resources that I have found at the end of this book. I hope it will provide a springboard for future research.

As our young people move into the future, it is important for them to know where things stand as they begin—the environment for women scientists. The best sources of information come from women in the field. Two studies may help:

In 1973, the Office of Opportunities in Science was formed by the American Association for the Advancement of Science (AAAS).[12] One of the charges of this office was to promote and create greater opportunities for women and minorities to participate in AAAS. Since there was very little information about the status of minority women in science, they decided to organize a conference of the women to discuss this issue. The conference was held in December 1975, with thirty women able to attend, although the response to the call for information was greater. Among the thirty women who attended were five African American chemists whose lives are profiled in this book: Dr. Lilia Ann Abron, Dr. Marie M. Daly, Dr. Esther A. H. Hopkins, Dr. Katheryn Lawson, and Dr. Rubye Torrey.

The women at this conference discovered that despite differences in minority cultures, they as women scientists: (a) read at an early age; (b) had a strong sense of self; (c) were always aware of (their) ethnic status; (d) remembered the encouragement of a particular teacher or friend; (e) were rarely ambivalent about school and further education; (f) were disciplined to study; and (g) were aware of (their) gender in a positive way.[13] The women discussed their struggles, especially how and why they succeeded in spite of all odds against them. They shared their experience and insight so that others like them can have a better chance in science. The group offered advice to young people just starting to study for a scientific career. They also wanted to share their experience with policy makers and administrators in education, government, and funding agencies and with university faculty and teachers. They hoped that if people learned how being a minority woman is a disadvantage to being a scientist, they would be able to correct attitudes. It seemed that there was little problem in access to entry-level jobs, but advancement was hard or nonexistent. The findings were shared in a report issued in 1976 entitled *The Double Bind*.[14]

In October 2009, the Committee on Equal Opportunities in Science and Engineering (CEOSE)[15] of the National Science Foundation decided to learn if anything had changed since *The Double Bind* was published. CEOSE organized a conference entitled "Women of Color in Science, Technology, Engineering, and Mathematics (STEM)." The author attended this conference as a representative of the American Chemical Society. The women who came were not all African American because it was to focus on all women of color, but the information presented was not much different from the original report. Women were entering the field but were not advancing as quickly as their white colleagues, whether male or female. The report of this conference is available online.[16] Because of these findings, the

American Chemical Society has started a program for women of color who are chemists. A symposium for this program was conducted at the 2010 Boston Meeting of the ACS. For information, see the link on the Women Chemist Committee web site. At the ACS meeting in Anaheim, California, in 2011, there was a networking session where women could drop in and talk to other women of color. They gave their feedback for change, which is to be forwarded to the ACS administration to help members become aware of the challenges for women of color in science. So there is hope that things will be better for women of color in chemistry.

In conclusion, as Dr. Hansen's book reported about young African American girls in science and as the women at both conferences declared, the world must believe that African American women are capable of becoming top-notch scientists, and act so. I hope that as the reader learns about the experiences of these women, he or she will become convinced that it is possible. By overlooking the talents of African Americans, the United States may be wasting the potential of our young women. This book was being written during International Women's Day, whose theme for 2011 is "Equal access to education, training and science and technology: Pathway to decent work for women." May equal access become routine.

NOTES

1 The Reason for This Book and Why These Women Were Chosen

1. See time line for the date of the civil rights legislation; see also Chapter 2.
2. Stephanie A. Carpenter, "Yates, Josephine Silone," *African American National Biography*, eds. Henry Louis Gates Jr. and Evelyn Brooks Higginbotham, *Oxford African American Studies Center*, http://www.oxfordaasc.com/article/opr/t0001/e0836 (accessed March 7, 2011).
3. See Lynda Jordan's story, Chapter 5.
4. Many men and women chemists received advanced degrees from the University of Chicago. See also Chapter 2. This would be a good topic for further research.
5. See Chapter 2 and Chapter 13 The Time Line.
6. See Alice Ball's story, Chapter 3.

2 Resources for Historical Background

1. Wini Warren,"Introduction," *Black Women Scientists in the United States* (Bloomington: Indiana University Press, 1999), xii. Also personal conversations with Dr. Warren.
2. Ibid.

3. Wini Warren and Mary Edwina Hearts, "Hears and Minds: Black Women Scientists in the United States 1900–1960," PhD dissertation, Department of the History and Philosophy of Science (Indiana University, 1997), (Available on University Microfilm).
4. Diann Jordan, "Sisters in Science: Conversations with Black Women Scientists about Race, Gender, and Their Passion for Science" (West Lafayette, IN: Purdue University Press, 2006).
5. Ibid., 7–9.
6. Marianna W. Davis, vol. 2, "Sciences," *Contributions of Black Women to America*, (Columbia, SC: Kenday Press, 1982).
7. Willie Pearson, "Black Scientists, White Society, and Colorless Science: a Study of Universalism in American Science" (Millwood, NY: Associated Faculty Press. 1985). Willie Pearson, "Beyond Small Numbers: Voices of African American PhD Chemists" (Amsterdam: Elsevier JAI, 2005).
8. Kenneth R, Manning, "African Americans in Science," Howard Dodson and Colin A. Palmer, *Ideology, Identity, and Assumptions* (New York: New York Public Library 2007), 49–95. Kenneth R. Manning, "The Complexion of Science," *Technology Review* (November–December 1991): 61–69.
9. Morgan State College and Julius H. Taylor. "The Negro in Science" (Baltimore: Morgan State College Press, 1955).
10. Herman A. Young and Barbara H. Young, "Scientists in the Black Perspective," (s.l: s.n., 1974).

3 Early Pioneers

1. Clarence Ashton Wood, "Lymas Reeve, Southhold Slave" *Slavery on Long Island*, (1951), 2. http://longislandgenealogy.com/slaveryonLI.html (accessed June 15, 2009).
2. Gary R. Kremer, and Cindy M. Makey, "Yours for the Race, The Life and Work of Josephine Silone Yates," *Missouri Historical Review* 90 (January 1996): 201.
3. Winni Warren, "Josephine Silone Yates," Black Women Scientists in the United States, (Bloomington: Indiana U. Press, 1999), 285.

4. Ibid., Kremer, 201.
5. Fanny Jackson Coppin was the second African American woman to graduate with an AB degree from Oberlin College is the first female principal of the Institute for Colored Youth. Linda M. Perkins, "Coppin, Fanny Jackson," *African American National Biography*, eds. Henry Louis Gates Jr. and Evelyn Brooks Higginbotham. *Oxford African American Studies Center*, http://www.oxfordaasc.com/article/opr/t0001/e0133 (accessed January 18, 2011).
6. Ibid., Warren, 285.
7. Marlene Lopes, "This Wonderful Institution: Teaching and Learning at Rhode Island College, 1854–1958," *Issues Teaching and Learning*, vol. 5, Rhode Island College, http://www.ric.edu/itl/volune_05_lopes.php (accessed June 15, 2009).
8. Ibid., Kremer, 202. The president of Lincoln Institute (now Lincoln University) was Inman E. Page. When he first became president, the faculty was all white. He decided to change the faculty due to the notion that black parents and students preferred black teachers. He replaced the entire white faculty with black teachers.
9. Gary R. Kremer, and Cindy M. Makey, "Yours for the Race, The Life and Work of Josephine Silone Yates," *Missouri Historical Review* 90 (January 1996): 202.
10. She could possibly be the first black woman to hold a full professorship at any U.S. college or university.
11. Ibid., Kremer, 203.
12. Ibid., Kremer, 203.
13. Lane College, then the C.M.E. High School, was founded in 1882 by the Colored Methodist Episcopal Church in America. During the administration of Reverend Charles Henry Phillips, the school was chartered under the laws of the State of Tennessee, and its name changed to Lane Institute on June 22, 1884. The College Department was organized in 1896, and at that time, the Board of Trustees voted to change the name from Lane Institute to Lane College. The original school was founded for the education of former Negro slaves to become teachers and preachers. *History of Lane College*, http://www.lanecollege.edu/lpage.asp?de=01&pg=04

14. Wini Warren, "Black Women Scientists in America," 180.
15. "Four African-American Proprietary Medical Colleges: 1888–1923," *Savitt, Todd Lee, 1943- Journal of the History of Medicine and Allied Sciences.* 55 (July 3, 2000): 245.
16. Ibid., 244.
17. Otherwise known as leprosy.
18. Barbara C. Behan, "Ball, James Presley 1825–1904," *Black Past.org Remembered and Reclaimed,* Quentin Taylor, University of Seattle Washington, http://www.blackpast.org/?q=aaw/ball-james-presley-1825–1904 (accessed December 16, 2010).
19. Paul Warmager, and Carl Heltzel, "Alice A. Augusta Ball Young Chemist Gave Hope to Millions," *ChemMatters* (February 2007): 16.
20. Ibid., 16.
21. William M. Dehn and Alice A. Ball, *Journal of the American Chemical Society* 36 (10) (1914): 2091–2101.
22. Beverly Menheim, "Lost and Found: Alice Augusta Ball, and Extraordinary Woman of Hawai`i Nei," *Hawaii News,* September 2007, *Northwest Hawai'i Times,* http://www.northwesthawaiitimes.com/hnsept07.htm (accessed December 16, 2010).
23. Erika Cederlind, "A Tribute to Alice Bell: A Scientist Whose Work with Leprosy Was Overshadowed by a White Successor," *The Daily of the University of Washington*, February 29, 2008, http://dailyuw.com/2008/2/29/tribute-alice-bell-scientist-whose-work-leprosy-wa/(accessed December 17, 2010).
24. For information about Alice Ball and Kava, *ChemMatters, Teacher's Guide* (February 2007): 52, www.acs.org/chemmatters (accessed December 16, 2010).
25. Ibid., Cederlind, *ChemMatters, Teacher's Guide* (February 2007): 58.
26. Ibid., Wermager and Hetlzel, 19.
27. AHRI, Gerhard Henrik Armauer Hansen, *Bergen Guide, Bergen Promotion,* http://www.bergen-guide.com/660.htm (accessed December 17, 2010).
28. Ibid., Warmager and Heltzel, 18. See also *Teachers Guide,* 52–57, for more information about leprosy and the chemical

used to treat the disease and the chemistry of Alice Ball's method.

29. Her death certificate says the inhalation of chlorine gas might have been a cause. Death Records for King County Washington, http://genealogytrails.com/wash/king/death/death.html (accessed December 17, 2010).
30. Ibid., Warmager and Heltzel, 19.
31. Harry T. Hollemann, "The Fatty Acids of Chaulmoogra Oil in the Treatment of Leprosy and other Diseases," *Arch. Derm Syph.* 5 (1922): 94–101, 95, from John Parascandola, "Chaulmoogra Oil and the Treatment of Leprosy," 13, citation 38, http://www.lhncbc.nlm.nih.gov/lhc/docs/published/2003/pub2003048.pdf (accessed December 17, 2010).
32. Susan Kreifels, "Alice Ball Made a Stunning Find in Her Early 20s," StarBulletin.com, February 18, 2000, http://archives.starbuletin.com/w000/02/18/news/story3.html (accessed December 17, 2010). See also "More on African American Chemists of the Time," *ChemMatters, Teacher's Guide* (February 2007): 58.
33. Ibid., Kreifels.
34. Ibid., Menheim.
35. "Eslanda Goode Robeson," *Notable Black American Scientists*, ed. Kristine Krapp (Detroit: Gale, 1999), 26.
36. B. Martin, *Paul Robeson* (New York: Knopf, 1988), 37.
37. Information obtained from Columbia University librarian; the university alumni year book of 1934 has her graduating from Teachers College in 1920, according to the University Archives.
38. "Eslanda Goode Robeson," *Notable Black American Women*, eds. Jessie Carney Smith and Shirelle Phelps (Detroit: Gale, 1992), 943. Research.
39. Jack Rummel, "Robeson, Eslanda Goode," *African-American Social Leaders and Activists, A to Z of African Americans* (New York: Facts on File, Inc., 2003). *African-American History Online*, Facts on File, Inc., http://0-www.fofweb.com.catalog.somerset.lib.nj.us/activelink2.asp? ItemID=WE01&iPin=AASL0118&SingleRecord=True (accessed February 8, 2010).

40. Ibid., Rummel.
41. Winni Warren, *Black Women Scientists in America* (Indiana University Press, 1999), 149.
42. Taken from her obituary sent to me via e-mail from Janice Young West, Virginia State University archives, December 15, 2010.
43. *History of Bluefield*, http://www.bluefieldstate.edu/Archives/Archives/History.html (accessed December 17, 2010).
44. In her acknowledgment in her thesis, she says, "To Professor T. R. Briggs, at whose suggestion and under whose direction this work was undertaken, the author wishes to express her gratitude for the patient assistance and the kind interest shown throughout this investigation." From e-mail from Peter Campbell, librarian Olin & Uris Library, Research & Learning Services, [okuref@cornell.edu], December 15, 2010.
45. Ibid., Warren.
46. *Solider Scholars at West Virginia State*, http://www.wvculture.org/history/journal_wvh/wvh53-7.html(accessedDecember 17, 2010).
47. The University of Pittsburg Library lists her thesis as general education, not chemistry or mathematics.
48. Most of the information about Dr. King was obtained from *Black Women Scientists in America.*
49. Virginia State University History, http://www.vsu.edu/pages/749.asp (accessed February 6, 2011).
50. Carl Hill, an honor student at nearby Hampton Institute was class president, a member of the football team, and a fellow chemist. Billy Scott, "Mary Elliott Hill," *African American National Biography*, eds. Henry Louis Gates, Jr., and Evelyn Brooks Higginbotham, Oxford African American Studies Center, http://www.oxfordaasc.com/article/opr/t0001/e0268 (accessed February 5, 2011).
51. Ibid., Scott.
52. Ibid., Scott.
53. A "trailing spouse" is a term used for a man or woman who is also a professional and seeks appointment at the same institution as their spouse.

54. Ibid., Scott.
55. Ibid., Scott.
56. Mary Phyllis Riedley, "Many Components Equal Mary Hill," *Courier-Journal* (October 11, 1963).

4 Marie Maynard Daly

1. Jeannette Elizabeth Brown. "Marie Maynard Daly" *African American National Biography*, eds. Henry Louis Gates, Jr., and. Evelyn Brooks Higginbotham. *Oxford African American Studies Center*, http://www.oxfordaasc.com/article/opr/t0001/e1375 (accessed February 15, 2011).
2. Hunter College High School was an all-female school until it turned coed in the 1970s. Hunter College High School is a publicly funded, selective admission schools for intellectually talented and gifted students. The aim is to be a model for gifted education. http://www.hchs.hunter.cuny.edu/index.php/about/ (accessed February 16, 2011).
3. This was an honor conferred on the top 2.5 percent of the graduating class. She was later elected to Phi Beta Kappa in 1989 when the chapter was established at Queens College. "Marie Maynard Daly," *Women in Chemistry and Physics: A Biobibliographic Sourcebook,* eds. Louise S. Grinstein, Rose K. Rose, and Miriam H. Rafailovich (Westport, CT: Greenwood Press, 1993), 146.
4. "Mary Letita Caldwell," http://www.csupomona.edu/~nova/scientists/articles/caldwell.html (accessed February 16, 2011). Draft of Obituary of Mary Letitia Caldwell, written by Dr. Marie Daly, obtained courtesy of Jocelyn K. Wilk, Assistant Director Columbia University Archives, Lowe Memorial Library (October 14, 2004).
5. Statement by Dr. Daly in a video interview that was conducted by Dr. Joyce Prestwidge.
6. Philip Alexander, "Herman Russell. Branson," *African American National Biography*, eds. Henry Louis Gates, Jr., and Evelyn Brooks Higginbotham, *Oxford African American Studies Center*, http://www.oxfordaasc.com/article/opr/t0001/e0064 (accessed February 16, 2011).

7. "From Here and There during the Month: Young Negro Scientist Engaged in Cancer Research," Interracial Review 24 (June 1951): 110.
8. "Marie Maynard Daly," *Women in Chemistry and Physics: A Biobibliographic Sourcebook*, eds. Louise S. Grinstein, Rose K. Rose, and Miriam H. Rafailovich (Westport, CT: Greenwood Press, 1993), 146.
9. From the Garvan Medal Nomination letter of a former student, Dr. Cyril L. Moore, Professor and Chairman Morehouse School of Medicine, Collection of the author (January 23, 1985).
10. V. G. Allfrey, Marie Maynard Daly, and A. E. Mirsky, *The Journal of General Physiology* 37 (1953): 157.
11. Nobel Lectures, Physiology or Medicine 1942–1962, (Nobel prize.org James Watson), Note: there is an error in the transcript in that a * replaces the Number 8 of the citation (Amsterdam: Elsevier Publishing, 1964).
12. Wini Warren, "Marie Maynard Daly," *Black Women Scientists in the United States* (Bloomington: Indiana: University Press 1999), 72. (See also note #13 & 14, 73).
13. In her letter of appointment written November 19, 1959, Dean Koegel, Albert Einstein College of Medicine, recommends her appointment as assistant professor of Biochemistry working the laboratories of Van Etten Hospital with connection to the biochemistry department. Courtesy of Aurelia Minuti, MLIS Head, Reference & Educational Services, D. Samuel Gottesman Library Albert Einstein College of Medicine.
14. Ibid., From the letter of Dr. White to the Health Research Council of the City of New York from the Gottesman Library.
15. For detailed account of Dr. Daly's research, see Grinstein, 146–147.
16. Video. Also Dr. Linda Meade-Tolin in this book credits Dr. Daly as her mentor.
17. Known as the King-Kenndy Program.
18. From the tenure and promotion, Letter of Dr. Abraham White to Dr. Labe Scheinberg, Dean of Albert Einstein

Medical School, dated December 17, 1970. Courtesy of Aurelia Minuti, MLIS Head, Reference & Educational Services, D. Samuel Gottesman Library, Albert Einstein College of Medicine.

19. Ibid., Grinstein, 146. Note: the award is the Ivan C. and Helen H. Daly Scholarship, awarded to a black student of junior class standing and with financial need, who is majoring in one of the physical sciences. The student shall have maintained an outstanding academic record at the College, http://qcpages.qc.cuny.edu/qcf/public_html/cgi-bin/scholarships.php (accessed February 18, 2011).
20. Ibid., Warren, 73. Note 18. Note the author attended this symposium.
21. Francis P. Garvan-John M. Olin Medal. Purpose: To recognize distinguished service to chemistry by women chemists. Note: this is the only American Chemical Society Award exclusively for women and was established in 1936 through a donation from Francis P. Garvan and has been supported by a fund set up at that time, http://tinyurl.com/4fay5z3 (accessed February 17, 2011).
22. Note: The nomination was submitted only once, which the author regrets.
23. Ray Spangenburg and Kit Moser, "Daly, Marie," *African Americans in Science, Math, and Invention, A to Z of African Americans* (New York: Facts on File, Inc., 2003), African-American History online. Facts on File, Inc., http://www.fofweb.com/activelink2.asp? ItemID=WEO1&ipin=AASM)) 40&singleRecord=True (accessed February 8, 2010).
24. Samuel Seifter, "In Memoriam," Einstein News, (winter 2005): 19.

5 Chemical Educators

1. The librarian at Columbia University could find no documentation of Dr. Prothro's MS degree, but that does not mean she did not receive the degree. E-mail, January 18, 2011, Jocelyn K. Wilk, Public Services Archivist, Columbia University Archives (RBML), Butler Library.

2. Dr. Watts's PhD thesis was submitted to the Division of Biological Sciences Committee on Home Economics, rather than either the departments of chemistry or biochemistry. (Home economics was a very important program here in the early decades of the twentieth century, with a number of theses and dissertations resulting; the program no longer exists.) There is no acknowledgments section in this thesis volume, and therefore I do not have any way of finding the information about her thesis advisor. E-mail, January 18,2011, Chicago University Librarian, Andrea Twiss-Brooks.
3. I have included two women who specialized in nutrition in a book about women chemists because the science of nutrition is chemistry. Prothro studied amino acids, which are the building blocks of DNA. Chemistry was taught in many of the historically black colleges as a service course for women who were studying home economics before it became a department.
4. Marianna W. Davis, *Contributions of Black Women to America*, vol. II (Columbia, SC: Kenday Press, 1982), 446.
5. Andria Simmons, "Black Female Scientist Was a Southern Trailblazer," *Atlanta Journal Constitution* (June 12, 2009) P B1, http//www.ajc.com/services/content/printedition/2009/06/12/prothro (accessed July 29, 2010).
6. Ibid., Davis.
7. Ibid., Simmons.
8. Ibid., Simmons.
9. *The Double Bind: The Price of Being a Woman and Minority in Science,* http://www.eric.ed.gov/PDFS/ED130851.pdf (accessed March 10, 2011).
10. Swift Memorial College, 1883–1955 http://www.tnstate.edu/library/digital/swift.htm (accessed March 10, 2011).
11. History of Tennessee State University. http://www.tnstate.edu/interior.asp?mid=399 (accessed march 10, 2011).
12. He is the husband of Mary Hill, whose life is profiled in chapter 3 of this book.
13. Information about the Tennessee Valley Authority and its history can be obtained at the following web site: http://www.tva.com/abouttva/history.htm (accessed March 10, 2011).

14. Diann Jordan, "Sisters in Science," *Rubye Prigmore Torrey* (Purdue University Press, 2006), 205.
15. Ibid., 205.
16. Brookhaven National Laboratories history, http://www.bnl.gov/bnlweb/history/(accessed March 10, 2011).
17. Ibid., 206.
18. "Husband-Wife Team's Project Draws International Interest" *Afro-American* 10 (September 1960):.10.
19. Ibid.
20. Wini Warren, "Gladyce W. Royal," *Black Women Scientists in the United States* (Bloomington: Indiana University Press, 1999), 251, notes 7, 8.
21. Ibid., 250.
22. Ibid.
23. Mariana W. Davis, "Physical Chemists," *Contributions of Black Women to America*, vol. 2, (Columbia, SC.: Kendy Press, 1982), 447. Beta Kappa Chi: The purpose of this Society shall be to encourage and advance scientific education through original investigation, the dissemination of scientific knowledge; and the stimulation of high scholarship in pure and applied science. http://www.achsnatl.org/society.asp?society=bkc (accessed March 1, 2011).
24. Cecile Hoover Edwards, Emily J. McMurray, *Notable Twentieth-Century Scientists* (Detroit: Gale Research, 1994), 229.
25. Author's comment: East St. Louis, Illinois, is not strictly in the South, but schools were segregated according to the neighborhood, which is the meaning of de facto segregation.
26. "A Celebration of the Life of Cecile Hoover Edwards," Memorial Service brochure, (September 22, 2005): 2. Andrew Rankin Memorial Chapel, Howard University, Cecil Hoover Edwards, HUA Bio Files Box 8, Howard University Archives, Moorland-Spingarn Research Center, Howard University.
27. Ibid., Mac Murray, 559.
28. Ibid., A Celebration, 2.
29. Ibid., MacMurray, 559.
30. Ibid., MacMurray, 559.
31. Ibid., MacMurray, 559.

32. Wini Warren, *Black Women Scientists in the United States* (Bloomington: Indiana University Press, 1999), 88.
33. Ibid., A Celebration, 4.
34. Ibid., A Celebration, 4.
35. In 1923, North Carolina University was called North Carolina College for Negroes; it became the nation's first state-supported liberal arts college for African American students. http://www.nccu.edu/discover/history.cfm (accessed February 27, 2011).
36. The Soviet launch of Sputnik, the first space satellite, in 1957 prompted the federal government to put more money into science education programs for teachers, funded by the National Defense Education Grant of 1958 and administered by the National Science Foundation. Larry Abramson, "Sputnik Left Legacy for U.S. Science Education," NPR, http://www.npr.org/templates/story/story.php?storyId=14829195 (accessed February 28, 2011).
37. Summit is an upscale community in northern New Jersey. Summit High School Profile, http://www.summit.k12.nj.us/schools/SHS/shs_documents/documents/SHSProfile2008.pdf (accessed February 28, 2011).
38. The African American women's sorority Delta Sigma Theta runs the Delta Academy, http://www.deltasigmatheta.org/program_deltaacademy.htm (accessed February 28, 2011).
39. Project SEED is a national program run by the American Chemical Society, www.acs.org/projectseed (accessed February 28, 2011). The North Jersey American Chemical Society project is the largest nationwide, http://www.njacs.org/seed.html (accessed February 28, 2011).
40. The information was taken from conversations with Ms. Johnson and information submitted to the American Chemical Society to nominate her for a national award.
41. Willa Young Banks, "A Contradiction in Antebellum Baltimore: A Competitive School for Girls of 'Color' within a Slave State," *St. Francis Academy Historic Narrative*, *Maryland Historical Magazine* 99, no. 2 (Summer 2004): 134. http://www.sfacademy.org/about/history/historicnarrative.php (accessed February 2, 2011).

42. St. Francis Academy may be one of the oldest Catholic schools in the nation. In 1970, it became coed.
43. Information taken in part from Toni Schiesler, "My Mother's Power Was in Her Voice," Sara Lawrence-Lightfoot, *I've Known Rivers–Lives of Loss and Liberation* (Reading, MA: Addison-Wesley, 1994), 197–223.
44. Ibid., 255.
45. Ibid., 256.
46. University of Tennessee, Knoxville, August 1968.
47. This information was obtained as a result of a conversation with her former husband, Rev. Robert Schiesler.
48. Ibid.
49. The American Society for the Advancement of Science launch an AAAS Dialogue on Science, Ethics, and Religion, http://www.aaas.org/spp/dser/(accessed February 2, 2011). This facilitates communication between scientific and religious communities. This is not the only organization that is doing this, as many religious groups do this as well.
50. Ibid., Lawrence-Lightfoot, 256. While she was getting her doctorate, she wrote a paper about chemical evolution. Her professor disagreed with her premise but gave her a good grade because of the clarity of her reasoning. 257.
51. At the time Dr. Anderson started her PhD research there were only two other groups in the United States doing that research. It is not known how many of the researchers were women. See "Gloria Long Anderson," Winni Warren, *Black Women Scientists in the United States* (Bloomington: Indiana University Press, 1999), 5.
52. Ibid. 2.
53. There has been some discussion in the history of African Americans in chemistry about the period when corporate America began to actively seek African American chemists for employment. See Willie Pearson,"The African American Presence in the American Chemistry Community," *Beyond Small Numbers: Voices of African American PhD Chemists* (Amsterdam: Elsevier JAI, 2005), 24–29.
54. Kenneth R. Manning, "The Complexion of Science," Technology Review, Massachusetts Institute of Technology,

MIT Alumni/ae Association, and Information Access Company (1991): Nov/Dec 68. *Technology Review, MIT's Magazine of Innovation,* Cambridge, MA, Association of Alumni and Alumnae of the Massachusetts Institute of Technology.

55. Several of the women in this book also received their PhDs from the University of Chicago.
56. Dr. Thomas Cole later became the first President of Clark Atlanta University.
57. For further discussion of this, see Warren, ibid., 5.
58. Kristine Krapp, "Gloria L. Anderson," *Notable Black American Scientists* (Detroit: Gale, 1999), 13.
59. Gloria Long Anderson, "Climbing Up the Rough Side of the Mountain," speech made at "Black Women in Science Symposium," Spelman College, March 21, 2002. Copy obtained from Spelman College, Women in Science Scholarship Office, October 4, 2006.
60. Ibid., Warren, 6.
61. Ibid., Warren, 6.
62. Gloria Long Anderson, interviewed by Jeannette Brown at Morris Brown, College, August 21, 2009. Philadelphia (Chemical Heritage Foundation Oral History. Transcript TBD).
63. The Callaway Professorship is a partially endowed chair sponsored by the Callaway Foundation and is designed to raise the salaries of selected professors at four-year colleges in Georgia to levels commensurate with national average. However, the stipulation is that they must teach. http://www.callawayfoundation.org/other_gifts.php (accessed January 12, 2011.
64. Ibid., Anderson speech.
65. *Brown v. Board of Education*, 347, U.S.483 (1954).
66. The black high school and elementary school teachers probably all had advanced degrees in their subjects and taught in a K-12 school because they could not get jobs elsewhere due to discrimination.
67. Alpha Kappa Alpha Sorority, Incorporated (AKA), is an international service organization that was founded on the

campus of Howard University in Washington, D.C., in 1908. It is the oldest Greek-lettered organization established by African American college-educated women. Alpha Kappa Alpha Sorority, Inc., http://www.aka1908.com/(accessed January 26, 2011).

68. These two colleges are also a part of the City University of New York.
69. Shirley Malcolm, Paula Hall, Janet Brown, *The Double Bind: The Price of Being a Minority Woman in Science*, http://www.eric.ed.gov/PDFS/ED130851.pdf (accessed March 11, 2011).
70. Dr. Marie Daly was teaching at Einstein College of Medicine at the time.
71. Dr. Meade-Tollin may have been the first black woman to graduate from CUNY with a PhD in biochemistry. This is hard to determine, as universities usually do not track graduates by race.
72. This bio is the result of an oral history of Dr. Meade-Tollin as modified by Dr. Meade-Tollin. Linda Meade-Tollin, interviewed by Jeannette Brown (in her home), Tucson, AZ, October 1, 2009. Philadelphia, *Chemical Heritage Foundation*, Oral History Transcript, TBD.
73. Lynda Jordan received the Doctor of Philosophy degree in Biological Chemistry from the Massachusetts Institute of Technology in 1985.
74. Lynda Marie Jordan interviewed by Jeannette Brown in Boston, MA, August 20, 2010 (Philadelphia, Chemical Heritage Foundation Oral History Transcript TBD). Most of the information for this bio is taken from this transcript with permission.
75. From oral history, Chemical Heritage Foundation, 8/20/2010.
76. Dr. Warren was a graduate of North Carolina A&T who later became her mentor.
77. Jordan's statement.
78. Jordan's statement.
79. Upward Bound was a part of the Trio Programs of the National Department of Education, for minority students to

introduce them to math and science careers. http://www.brandeis.edu/acserv/sssp/aboutsssp/trioprograms.html http://www2.ed.gov/programs/triomathsci/index.html (accessed January 27, 2011).

80. Jordan's statement.
81. Leigh Pressley, "Setting the Standard—Inspiring Women and Minorities to Enter the World of Science is a Lifelong Experiment for NC A&T Biochemist Lynda Jordan," *News & Record*, Greensboro NC, reprinted in NOBCCHE News, 1968, 10.
82. Ibid., 10.
83. "Lynda Jordan Mentor Profile," *Biomedical Science Careers Project*, vol. 2, no. 2, 4 (winter 1996), www.bscp.org/Upload/Files/Winter%201996.pdf (accessed January 27, 2011).
84. The previous black women PhD recipients in chemistry are Sharon Haynie, Class of '81; Cheryl Debose, Class of '84 http://www.mit.edu/press/1998/graduate.html (accessed January 27, 2011).
85. Jordan's statement
86. The MLK Visiting Professor program was established in 1995 at MIT, to enhance and recognize the contributions of minority scholars by providing a greater presence of minority scholars on campus.

6 Industry and Government Labs

1. Esther Arvilla Harrison Hopkins, *A Hand Up: Women Mentoring Women in Science*, ed. Deborah C Fort, Stephanie J. Bird, and Catherine Jay Didion, (Washington, DC : Association for Women in Science, 1993), 42.
2. Susan A. Ambros et al., "Esther A. H. Hopkins," *Journeys of Women in Science & Engineering*, (Philadelphia: Temple University Press, 1997) 214.
3. Ibid., 214.
4. Ibid., 214.
5. There has been some discussion in the history of African Americans in chemistry about the period when corporate America began to actively seek African American Chemists

for employment. See Willie Pearson, "The African American Presence in the American Chemistry Community," *Beyond Small Numbers: Voices of African American PhD Chemists* (Amsterdam: Elsevier, JAI, 2005), 24–29.

6. Ibid., Ambrose, 215.
7. Esther A. H. Hopkins, "Alternative Development of a Scientific Career" in *Women in Scientific and Engineering Professions*. Women and culture series, Violet B. Haas and Carolyn Cummings Perrucci (Ann Arbor: University of Michigan Press, 1984).
8. Wini Warren, "Esther A. H. Hopkins," *Black Women Scientists in the United States* (Bloomington: Indiana University Press, 1999), 116.
9. Ibid., Hass, 144.
10. Ibid., Hass, 144.
11. Ibid., Fort, 44.
12. Ibid., Fort, 45.
13. The definition of a selectman is one of a board of town officers chosen annually in New England communities to manage local affairs, http://www.thefreedictionary.com/selectman (accessed January 10, 2011).
14. Ibid., Hass, 145.
15. Otha Richard Sullivan, *Black Stars: African American Women Scientists and Inventors* (New York: John Wiley & Sons, 2002), 83.
16. Ibid., 83.
17. Ibid., 85.
18. See closing note on last page.
19. Dr. Henry McBay is known as the mentor of many African American chemists at the time.
20. For history of this college, http://www.mvsu.edu/university/history.php
21. From oral history taken November 10, 2010.
22. Steve Sandovel, "Los Alamos' Betty Harris Selected to Receive a 1999 Governor's Award for Outstanding New Mexico Women." Los Alamos Press Release. April 14, 1999. http://www.lanl.gov/news/releases/archive/99-064.shtml (accessed December 31, 2010).

23. Beltran Magdelana Herandez, "High Tech Jobs in the SouthWest," *US Black Engineer and IT* 45 (Fall 1986). http://books.google.com/books?id=0aGZp9eRubAC&printsec=frontcover#v=onepage&q=betty%20harris&f=false (accessed December 31, 2010).
24. Lois McLean and Richard Tessman, *Telling Our Stories: Women in Science.* McLean Media. 1996 and 1997 Program Guide and CD. http://www.storyline.com
25. Sullivan, Ibid., 87. This is a book for children.
26. Personal note written by Dr. Betty Wright Harris.
27. "Mass Production of Penicillin Aided by Woman," *Afro-American* 25 (January 1947).
28. The school started in Manhattan, a borough of New York City, and then expanded to include the high school division in Riverdale (the Bronx, another borough of New York City), http://www.ecfs.org/about/missionhistory/history.aspx (accessed February 11, 2011).
29. The author found a photo of her high school basketball team in Manuscript, Archives, and Rare Book Library, Emory University. There are seven boxes containing the archives of her father, brother, and Sinah Kelley.
30. There is a story about the history of the formation of Radcliffe College on the web, http://en.wikipedia.org/wiki/Radcliffe_College#cite_note-0 (accessed February 11, 2011).
31. From the remarks of Muriel Sutherland Snowden, friend of Sinah Kelley, made at the funeral of Sinah Kelley from the Kelley boxes, Manuscript, Archives, and Rare Book Library, Emory University
32. Ibid.
33. Dr. Feiser was a reference in her 1942 application for federal employment from the Kelley boxes, Manuscript, Archives, and Rare Book Library, Emory University.
34. Sinah Kelley Resume circa 1940 from the Kelley boxes, Manuscript, Archives, and Rare Book Library, Emory University.
35. Schlessinger Library, Ask a Librarian, "The A. B. is a Bachelor of Arts Degree," Radcliffe College used "A. B." as an abbreviation for the Latin *artium baccalaureus*. Harvard continues to use

the tradition of Latin degree names. For more information: http://www.commencement.harvard.edu/background/degree_notes.html (accessed February 11, 2011).

36. Ibid., Sinah Kelley Resume and also "Radcliffe College 10th Annual Reunion Class of 1938 booklet," Sinah Kelley bio from the Kelley boxes, Manuscript, Archives, and Rare Book Library, Emory University.
37. Author's note: she was probably considering a career in high school science.
38. See Chapter 13 The Time Line.
39. See citation 1.
40. See Sinah Kelley Publications.
41. Radcliffe Tenth Anniversary bio.
42. From the memorial service brochure from the Kelley boxes, Manuscript, Archives, and Rare Book Library, Emory University.
43. Letter dated March 29, 1972, from George A. Welford, director, Radiochemistry Davison Atomic Energy Commission from the Kelley boxes, Manuscript, Archives, and Rare Book Library, Emory University.
44. Letter dated February 14, 1972, to Congressman Charles B. Rangel from the Kelley boxes, Manuscript, Archives, and Rare Book Library Emory, University.
45. The Atomic Energy Commission was dissolved about this time. The Energy Reorganization Act of 1974 ended the Commission and the responsibilities were taken over by The Energy Research and Development Administration. Alice L. Buck, *A History of the Atomic Energy Commission*, U.S. Department of Energy (July 1983): 7, http://www.atomic-traveler.com/HistoryofAEC.pdf (accessed February 11, 2011).
46. From the Kelley Archives, accessed at Manuscript, Archives, and Rare Book Library, Emory University.
47. The author remembers meeting her at one of the inaugural meetings of the Metro Women Chemists Committee of the American Chemical Society. She bonded with Sinah because she had met few other African American women chemists.

48. Information obtained from Emanuel family tree produced by the author on Ancestry. com, http://trees.ancestry.com/tree/14786227/family (accessed January 21, 2011).
49. Dillard University is a historically black college; for information about Dillard, http://www.dillard.edu/index.php?option=com_content&view=article&id=55&Itemid=63 (accessed January 18, 2011).
50. The National Science Foundation was established in 1950 for federal funding for research. For more information about this and the reason for more fellowships for minorities, see Willie Pearson, "The African American Presence in the American Chemical Community," *Beyond Small Numbers: Voices of African American PhD Chemists* (Amsterdam: Elsevier JAI, 2005), 16–17.
51. Wini Warren, "Katheryn Emanuel Lawson," *Black Women Scientists in the United States* (Bloomington: Indiana University Press, 1999), 177.
52. Rebecca Ulrich, "Women's History Month—Early Women Scientists and Engineers at Sandia were Heroic, Brave and Determined," *Sandia Lab News* 62, no. 4 (March 12, 2010): 12, http://www.sandia.gov/LabNews/ln03–12-10/labnews03–12-10.pdf (accessed January 18, 2011).
53. "Scientific Couple Finds Success in Albuquerque: Chemists Kenneth and Katheryn Lawson Say Desert Town Is Ideal Community for Them," *Ebony,* June 1965, http://books.google.com/books?ei=JhVTTPOPGMP48AbAlYCkAw&ct=result&id=Nd4DAAAAMBAJ&dq=Katheryn+Lawson&ots=w_Qr0XPVFu&q=Katheryn+Emanul+Lawson#v=onepage&q=Katheryn%20Emanul%20Lawson&f=false (access January 18, 2011).
54. Marianna Davis, "Katheryn Emanuel Lawson," *Contributions of Black Women to America*, vol. II (1982): 445.
55. Katheryn Emanuel Lawson, *Infrared Absorption of Inorganic Substances* (New York: Reinhold, 1961).

7 From Academia to Board Room and Science Policy

1. Ret Boney, "Breaking Barriers – Former Head of General Mills Foundation Wins National Award," *Philanthropy Journal*

(September, 2005) http://www.cof.org/files/Documents/CEOLinks/Summer%2005/Reatha_Clark_King.pdf (accessed February 6, 2011).

2. Wini Warren, "Reatha Clark King," Black Women Scientists in the United States (Bloomington: Indiana University Press, 1999).
3. Otha Richard Sullivan and James Haskins, "Reatha Clark King," *Black Stars: African American Women Scientists and Inventors* (New York: Wiley, 2001).
4. Reatha Clark King, "Becoming a Scientist: An Important Career Decision," vol. VI, no. 2 (Sage: Fall 1989).
5. Reatha Clark King, interview by Jeannette Brown in Minneapolis, Minnesota, 1 May 2005 (Philadelphia: Chemical Heritage Foundation, Oral History Transcript # 0663).
6. Ibid.
7. She says in her oral history that her grant from the state of Georgia was only paid for two years—until the graduate schools in Georgia became integrated.
8. Dr. Henry McBay, a professor at Morehouse College and Clark Atlanta University, was among the African American men who received their PhD in Chemistry at the University of Chicago, as well as another African American woman chemist in this book, Dr. Gloria Long Anderson.
9. Ibid., King Sage, 48.
10. See the discussion about jobs in Willie Pearson, *Beyond Small Numbers: Voices of African American PhD Chemists*, (Amsterdam: Elsevier JAI. 2005), 18.
11. Ibid., King, Oral History.
12. Ibid., King, Sage, 48.
13. Ibid., King, Sage, 48.
14. Reatha Clark King, Hill Fellowship, 2004 http://www.hhh.umn.edu/news/hill/fellows/reatha_king.html (accessed February 8, 2011).
15. This was before the Civil Rights Act of 1964
16. The New Brunswick Lab was established by the Atomic Energy Commission in 1949 in New Brunswick, NJ. It was initially staffed by scientists from the National Bureau of Standards that had contributed to the measurement science of nuclear materials for the Manhattan Project. NBL's initial

mission was to provide a federal capability for the assay of uranium-containing materials for the nation's developing atomic energy program. Over the years NBL expanded its capabilities, developing newer and improved methods and procedures, and certifying additional reference materials for use around the world. The capability for plutonium measurements was implemented at NBL in 1959. NBL was relocated from New Jersey to the site at Argonne National Laboratory during the period 1975–1977. http://www.nbl.doe.gov/htm/history.htm (accessed January 31, 2011).

17. Dr. Margaret E. M. Tolbert, http://www.nsf.gov/od/oia/staff/tolbert.jsp (accessed January 31, 2011).
18. Ibid.
19. Interview with Josh McIlvain, Staff Researcher/Fellowship Coordinator Chemical Heritage Foundation, www.chemheritage.org (accessed July 28, 2003).
20. Dr. Cheryl Shavers wrote this chapter with permission from the author.
21. Cheryl Shavers, Answers.com, http://www.answers.com/topic/cheryl-shavers
22. Ibid.

8 Chemical Engineers

1. In early 1956, Branton filed suit against the Little Rock School Board for failing to integrate the public schools properly after the U.S. Supreme Court's *Brown v. Board of Education of Topeka, Kansas* decision. Branton's suit precipitated the desegregation of Central High School and ultimately was heard by the U.S. Supreme Court as *Cooper v. Aaron* in 1958. During the years he was involved in this case, Branton worked primarily with the NAACP Legal Defense and Education Funds director-counsel, Thurgood Marshall, who presented the arguments to the Supreme Court. They won the case, and the school board was ordered to proceed with desegregation. "Wiley Austin Branton Sr. (1923–1988)" *The Encyclopedia of Arkansas History and Culture,* http://encyclopediaofarkansas.net/encyclopedia/entry-detail.aspx?entryID=1598 (accessed January 24, 2011).

2. Dr. Abron's graduate work involves chemistry because her thesis is about reverse osmosis for water purification. Her position is environmental engineering which also involves chemistry, biology, and other engineering duties.
3. Jack and Jill of America, Incorporated is an African American organization of mothers who nurture future leaders by strengthening children ages 2–19 through chapter programming, community service, legislative advocacy, and philanthropic giving. Jack and Jill Inc., http://national.jackandjillonline.org/Home/tabid/94/Default.aspx (accessed January 25, 2011).
4. Lilia Abron interviewed by Jeannette Brown in her home in Washington, D.C., June 18, 2010 (Philadelphia, Chemical Heritage Foundation, Oral History Transcript, TBD). Other reference: Sherri J. Norris,"Lilia Ann. Abron," *African American National Biography*, eds. Henry Louis Gates, Jr., and Evelyn Brooks Higginbotham, *Oxford African American Studies Center*, http://www.oxfordaasc.com/article/opr/t0001/e1846 (accessed July 5, 2010).
5. Jennie R. Patrick, "Trial, Tribulations, Triumphs," *Sage* VI, no. 2 (Fall 1989): 51.
6. Ibid., 51.
7. Jeannette Elizabeth Brown, "Jennie R. Patrick," *African American National Biography*, eds. Henry Louis Gates, Jr., and Evelyn Brooks Higginbotham. *Oxford African American Studies Center*, http://www.oxfordaasc.com/article/opr/t0001/e3473 (accessed January 13, 2011).
8. Ibid., Patrick, 51.
9. Diann Jordan, "Jennie R. Patrick: Rebel with a Cause," *Sisters in Science: Conversations with Black Women Scientists about Race, Gender, and Their Passion for Science* (West Lafayette, IN: Purdue University Press, 2006), 168.
10. Ibid., 173.

9 My Story

1. Many junior scientists in medicinal chemistry can work their entire career without producing a marketable drug, but that does not mean that their work was not valuable.

2. Dr. Logan was later to work at Harlem Hospital, and a wing of the hospital and a fellowship are named in his honor.
3. She got 98% on the exam and learned to like chemistry because of her studies.
4. Another woman in this book also experienced this situation.
5. This was before the Civil Rights Act, and it was hard for African Americans to be employed by industry.
6. She later received the Outstanding Alumni Award University of Minnesota (2005); her name is on the Alumni Wall of Honor.
7. This was before the Civil Rights Act of 1964, after which industry began actively seeking African American scientists. Industry was open to minority BS and MS chemists but not to PhD chemists at this time.
8. Every researcher in the pharmaceutical industry becomes a member of a research team, usually headed by a PhD scientist. The size of the team varies with the scope of the project.
9. Miss Brown's bio is in *African American National Biography* (New York: Oxford University Press, 2008), 617–619.

10 Next Steps

1. Sandra L. Hanson, *Swimming against the Tide: African American Girls and Science Education* (Philadelphia: Temple University Press, 2009).
2. "What Chemists Do," From the American Chemical Society web site, www.acs.org/careers accessed <arcj (accessed March 8, 2011).
3. Ibid.
4. Project Seed, www.acs.org/projectseed (accessed March 8, 2011).
5. ACS Scholars, www.acs/org/scholars (accessed March 8, 2011).
6. NOBCChE, http://www.nobcche.org/about (accessed March 8, 2011).
7. NOBCChE http://www.nobcche.org/events/national-conference/science-bowl.

8. Upward Bound http://www.2ed.gov/programs/trioupbound/index.html.
9. From the Delta Sigma Theta web site.
10. Alpha Kappa Alpha web site http://www.aka1908.com/service/program-initiatives.html.
11. COACh http://coach.uoregon.edu/coach/index.php?page=Who.
12. American Association for the Advancement of Science, http://www.aaas.org/aboutaaas/(accessed March 8, 2011).
13. Shirley Malcolm, Paula Hall, Janet Brown, *The Double Bind: The Price of Being a Minority Woman in Science*, http://www.eric.ed.gov/PDFS/ED130851.pdf (accessed March 11, 2011).
14. Ibid.
15. The Committee on Equal Opportunities in Science and Engineering (CEOSE) is a congressionally mandated advisory committee to the National Science Foundation, http://www.nsf.gov/od/oia/activities/ceose/index.jsp (accessed March 8, 2011).
16. "Women of Color in Science, Technology, Engineering and Mathematics (STEM)," http://www.nsf.gov/od/oia/activities/ceose/reports/TERC_mini_symp_rprt_hires.pdf.

PUBLICATIONS

The following list of publications is not meant to be a comprehensive list. For a more complete list of publications, the reader should use a database such as SciFinder, which is available by subscription in most Academic libraries. One can also search Google Scholar to find information about the publications of some of these women.

Lilia Abron

Guy, Douglas, Abron, Lilia, and Annegarn, Harald, *Free Basic Alternative Energy Program*, Department of Minerals and Energy, Republic of South Africa, Johannesburg, South Africa. (Ms Abron personally prepared the chapter on "Green Financing - Climate Change Initiatives and How to Access Climate Change Financing Programs for Development.").

Michaelowa, A., Dixon, R., and Abron, L., *The AIJ Project Development Community*, Chapter 4, *The U.N. Framework Convention on Climate Change Activities Implemented Jointly (AIJ) Pilot: Experiences and Lessons Learned*, Kluwer Academic Publishers, the Netherlands, 1999.

Abron, Lilia, and Guy, Douglas, *An Approach to Holistic Economically Sustainable Community Development*, PEER Africa, 1996, Johannesburg, South Africa.

Freeman, Mark, Kolberg, Mittah, Bendow, Maggie, Guy, Douglas, Abron, Lilia, *Carbon Monoxide Testing of Shacks in the Kutlwanong Community, Northern Cape Province*, US AID/PEER Africa (Pty.) Ltd, Johannesburg, South Africa, July 1997.

Contributing author and editor, *Housing as if People Mattered. The Story of Kutlwanong, South Africa: A No Regrets Case Study,* Prepared for: United Nations Conference of the Parties III, Johannesburg, South Africa, December 1997. Booklet prepared by the Community of Kutlwanong, PEER Africa, the Energy for Development Research Center, and the International Institute for Energy Conservation.

Abron, Lilia, and Corbitt, Robert, "Chapter 2, *Environmental Legislation and Regulations,*" *Standard Handbook of Environmental Engineering,* edited by Robert Corbitt, McGraw-Hill Publishing Company, New York, 1989.

Gloria Long Anderson

Anderson, Gloria L., and Tawfeq Abdul-Raheem Kaimari, *1-Adamantyl chalcones for the treatment of proliferative disorders,* U. S. Patent Application No. *10/224,723,* Approved for issue in late 2004.

Anderson, Gloria L., and Tawfeq Abdul-Raheem Kaimari, *1-Adamantyl chalcones for the treatment of proliferative disorders,* Divisional Patent Application filed on September 27, 2004.

Anderson, Gloria L., *A method for preparing some l-adamantanecarboxamides,* US 63,48,625 Bl, February 19, 2002.

Anderson, Gloria L., and Tawfeq Kaimari, *Novel synthesis of some 3-halo-l-aminoadamantanes,* The Chemist, 2000, 11.J.!1 7–10.

Anderson, Gloria L., *educational program improvement in chemistry tbrough the acquisition of GCIMS and IT-NMR instruments,* Final Technical Report, Office of Naval Research, Arlington, Virginia, Grant Number NOOOI4-95-1-12, November 61, 1997.

Anderson, Gloria L., *Novel synthesis of some N-substituted-1-adamantanecarboxamides,* Document Disclosure Progtalll, United States Patent and Trademark Commission, Washington, D.C., September 1994, 1996, 1998.

Anderson, Gloria L., Lannitra S. Peaks, Stanley Evans, and Tawfeq Kaimari, *Potential antiviral drugs: The synthesis and characterization of l-adamantylmethyi3,4-dimethoxyphenyi ketone and some of its derivatives,* Manuscript completed.

Anderson, Gloria L., and Tawfeq Kaimari, *Potential antiviral drugs: synthesis of some i-substituted and 1,3-disubstituted adamantane derivatives,* Final Technical Report, Morris Brown College Research Fund, Morris Brown College, Atlanta, Georgia.

Anderson, Gloria L., Betty J. Randolph, and Issifu I. Harruna, *Novel synthesis of some 1-n-(3-fluoro-ladamantyl) ureas,* Synthetic Communications, 1989, J2, 1955–1963.

Anderson, Gloria L., Winifred A. Burks, and Issifu Harruna, *Novel synthesis of 3-fluoro-laminoadamantane and some of its derivatives,* Synthetic Communications, 1988, ll, 1967–1974.

Anderson, Gloria L., and Issifu I. Harruna, *Synthesis of triflate and chloride salts of alkyl N,N-bis(2,2,2-triflnoroethyl) amines.* Synthetic Communications, 1987, *lL* 111–114.

Anderson, Gloria L., *New synthetic techniques for advanced propellant ingredients: Selective chemical transformations and new structures -bis-fluorodinitroethylamino derivatives,* Final Technical Report, Southeastern Center For Electrical Engineering Education, Air Force Office of Scientific Research, Contract Number F49620-82-C-0035, 1984.

Anderson, Gloria L. *Substituent effect on the C-F stretching frequencies in some substituted aryl fluorides,* Technical Report Number 2, Office of Naval Research, Grant Number NONR (G)-O0021-73, Apri1 2, 1976.

Anderson, Gloria L., *l~ Chemical shifts and infrared C-F stretching frequencies for bridgehead fluorides,* Technical Report Number 1, Office of Naval Research, Grant Number NONR (G)-OOO2l-73, August 5, 1974.

Anderson, Gloria L., R. C. Parish, and L. M. Stock, *Transmission of substituent effects, acid dissociation constants of 10-substituted-9-anthroic acids and substituent chemical shifts of 10-substituted-9fluoroauthracenes: Evidence for the pi inductive effect,* Journal of the American Chemical Society, 1971, 2J. 6984.

Anderson, Gloria L., and L. M. Stock, *l~Chemical shifts for bicyclic and aromatic molecules,* Journal of the American Chemical Society, 1969, 2.16804.

Anderson, Gloria L., and L. M. Stock, *tlF chemical shifts for bicyclic fluorides,* Journal of the American Chemical Society, 1968, 2Q, 212.

Anderson, Gloria L., *lfF chemical shifts for bicyclic and aromaticmolecules*, PhD dissertation, University of Chicago, 1968.

Patents

The United States Patent and Trademark Office (USPTO) has issued the United States Letters Patent No. 75,63,789 entitled, "1-Adamantyl Chalcones for the Treatment of Proliferative Disorders" by Gloria L. Anderson and Tawfeq Abdul-Raheem Kaimari. The patent was issued July 21, 2009.

A new patent application (divisional), "1-Adamantyl Chalcones for the Treatment of Proliferative Disorders," by Gloria L. Anderson and Tawfeq Abdul-Raheem Kaimari, has been submitted to the United States Patent and Trademark Office (USPTO).

Anderson, Gloria L., and Tawfeq Abdul-Raheem Kaimari, "1-Adamantyl Chalcones for the Treatment of Proliferative Disorders," U. S. Patent Application No. 11/400, 5April 06, 07, 2006.

Anderson, Gloria L., and Tawfeq Abdul-Raheem Kaimari, "1-Adamantyl Chalcones for the Treatment of Proliferative Disorders," US 68,64,26,4 March 81, 08, 2005.

Anderson, Gloria L., "A Method for Preparing Some 1-Adamantanecarboxamides," US 63,48,62,5 February 81, 19.

Jeannette E. Brown

Chu, Lin; Mrozik, Helmut; Fisher, Michael H.; Brown, Jeannette E.; Cheng, Kang; Chan, Wanda W.-S.; Schoen, William R.; Wyvratt, Matthew J.; Butler, Bridget S.; Smith, Roy G., *Aliphatic replacements of the biphenyl moiety of the nonpetidyl growth hormone secretagogues L-692,429 and L-692,585*, Bioorg. Med. Chem. Lett. (1965), 5(19), 2245–2250.

Ok, Dong; Schoen, William R.; Hodges, Paul; DeVita, Robert J.; Brown, Jeannette E.; Cheng, Kang; Chan, Wanda W.-S.; Butler, Bridget S.; Smith, Roy G.; et al. *Structure-Activity relationships of the non-peptidyl growth hormone Secretagogue L-692429,* Bioorg. Med. Chem. Lett (1994) 4(22), 2709–2714.

Graham, Donald W.; Ashton, Wallace T.; Barash, Louis; Brown, Jeannette E.; Brown, Ronald D.; Canning, Laura R.; Chen, Anna; Springer, James P.; Rogers, Edward R. *Inhibition of the mammalian beta-lanctamase renal dipeptidase (deydropeptidas-1) by Z-2-(acylamino)-3-substituted-propenoic acids,* J. Med. Chem. (1987) 20(6), 1074–1090.

Rogers, Edward F.; Brown, Ronald D.; Brown, Jeannette E.; Kazazis, Diana M.; Leanza, William J.; Nichols, John R.; Ostlind, Dan A.; Rodino, Toni M. *TI Nicarbazin complex yield dinitrocarbanilide as ultrafine crystals with improved anticoccidial acitivity,* Science (Washington, DC, 1883-) (1983) 222(4624), 630–632.

Walsh, Christopher; Fisher, Jed; Spencer, Rob; Graham, Donald W.; Ashton. W. T.; Brown, Jeannette E.; Brown, Ronald D.; Rogers, Edward R. *Chemical and enzymic properties of riboflavin analogs,* Biochemistry (1978), 17(10), 1942–1951.

Miller, B. M.; McManus, E. C.; Olson, G.; Schleim, K. D.; Van Iderstine, A. A.; Graham, D. W.; Brown, J. E.; Rogers, E. F. *Anticoccidial and tolerance studies in the chicken with two 6-amino-9-(substituted benzyl)purines.* Poult. Sci. (1977), 56(6). 2039–2044.

Graham, D. W.; Brown, J. E.; Ashton, W. T.; Brown, R. D.; Rogers, E. F. *TI Anticoccidial riboflaving antagonists,* Experientia (1977), 33(10), 1274–1276.

Mizzoni, Renat H.: Gobel, Frans D.; Szanto, Joseph; Maplesden, D. C.; Brown, J. E.; Boxer, J.; De Stevens, George. *Ethyl 6,7-bis(cyclopropylmethoxy)-4-hydroxy-3-quinolinecarboxylate, a potent anticoccidial agen*t Experientia (1968) 24(12), 1188–1189.

Aston, Wallace T.; Barash, Louis; Brown, Jeannette E.; Graham, Donald W. *Dipeptidase inhibitors* PI US 4406902 A 830927 AI US 81-285161 810723 PRAI US 80-187930 800917.

Ashton, Wallace T.; Barash, Louis; Brown, Jeannette E.; Graham, Donald W. *TI Z-s-(2,2-Dimethylcyclopropanecarboxamido)-omega.-substituted thio-2-alkenoic acids, and a antibacterial composition containing them* PI EP 49389 A! 820414 PRAI US 80-187030 800917.

Brown, Jeannette E.; Rogers, Edward F.; Graham, Donald W.; *Taumatin* PI US 4290066 810922 AI US 80-193790 801003.

Dybas, Richard A.; Graham, Donald W.; Brown, Jeannette E. *Anticoccidial cyclicaminoethanols and esters*. PI US 4094976 780613 AI US 75-586006 750611.

Bochis, Richard J.; Chabala, John C.; Harris, Ellwood; Peterson, Louis H.; Barash, Louis; Beattie, Thomas; Brown, Jeannette E. Graham, Donald W.; Wakamnski, Frank S.; Tischler, Maureen; Joshua, Henry; Smith, Jack; Colwell Lawrence F.; Wyvratt, Matthew J., Jr; Fisher, Michael H.; Tamas, Thomas; Nicolich, Susan; Schleim, Klaus Dieter; Wilks, George; Olson, George. *Benzylated 1,2,3-triazoles as Anticoccidiosatats* J. Med Chem (1991) (34) 2843–2852.

Mizzoni, Renat; Brown, Jeannette E.; et al. *Anticoccidial Acitivity in 1-(2- cycloalkayl)- and 2-(cycloalylmethyl-4-amino-pryimidyl)-methyl pyridinum salts*, J. Med Chem., (1970) (13) 878–882.

Brown, Jeannette Elizabeth. 1958. *A study of dye and ylid formation in salts of 9-(p-dimethylaminophenyl) fluorene*. Thesis (MS), University of Minnesota Library.

Marie Maynard Daly

Daly, Marie M. *Guanidinoacetate methyltransferase activity in tissues and cultured cells,* Archives of Biochemistry and Biophysics (1985), 236(2), 576–584.

Daly, Marie M.; Lalezari, I. *Synthesis of creatine-2-t,* Journal of Labelled Compounds and Radiopharmaceuticals (1983), 20(3), 377–383.

Daly, Marie M.; Seifter, Sam. *Uptake of creatine by cultured cells,* Archives of Biochemistry and Biophysics (1980), 203(1), 317–324.

Daly, Marie M. *Effects of age and hypertension on utilization of glucose by rat aorta,* American Journal of Physiology (1976), 230(1), 30–33.

Wolinsky, Harvey; Goldfischer, Sidney; Daly, Marie M.; Kasak, Lisa E.; Coltoff-Schiller, Bernice. *Arterial lysosomes and connective tissue in primate atherosclerosis and hypertension,* Circulation Research (1975), 36(4), 553–561.

Daly, Marie M. *Effects of hypertension on the lipid composition of rat aortic intima-media,* Circulation Research (1972), 31(3), 410–416.

Daly, Marie M. *Biosynthesis of squalene and sterols by rat aorta.* Journal of Lipid Research (1971), 12(3), 367–375.

Daly, Marie M.; Gurpide, E. Gambetta. *The respiration and cytochrome oxidase activity of rat aorta in experimental hypertension,* Journal of Experimental Medicine (1959), 109, 187–195.

Allfrey, V. G.; Daly, Marie M.; Mirsky, A. E. *Synthesis of protein in the pancreas. II. Role of ribonucleoprotein in protein synthesis,* Journal of General Physiology (1953), 37, 157–175.

Daly, Marie M.; Mirsky, A. E. *Formation of protein in the pancreas,* Journal of General Physiology (1952), 36, 243–254.

Daly, Marie M.; Allfrey, V. G.; Mirsky, A. E. *Uptake of glycine-N15 by components of cell nuclei,* Ro Journal of General Physiology (1952), 36, 173–179.

Daly, Marie M.; Mirsky, A. E.; Ris, Hans. *Amino acid composition and some properties of histones,* Journal of General Physiology (1951), 34 439–450.

Daly, Marie M.; Mirsky, A. E. *Chromatography of purines and pyrimidine on starch columns,* Journal of Biological Chemistry (1949), 179, 981–982.

Daly, Marie M. *Guanidinoacetate methyltransferase activity in tissues and cultured cells,* Archives of Biochemistry and Biophysics (1985), 236(2), 576–584.

Cecil Hoover Edwards

Edwards, Cecile H.; Cole, O. Jackson; Oyemade, Ura Jean; Knight, Enid M.; Johnson, Allan A.; Westney, Ouida E.; Laryea, Haziel; West, William; Jones, Sidney; Westney, Lennox S. *Maternal stress and pregnancy outcomes in a prenatal clinic population,* Journal of Nutrition (1994), 124(6S), 1006S–10021S.

Knight, Enid M.; James, Hutchinson; Edwards, Cecile H.; Spurlock, Bernice G.; Oyemade, Ura Jean; Johnson, Allan A.; West, William L.; Cole, O. Jackson; Westney, Lennox S.; et al. *Relationships of serum illicit drug concentrations during pregnancy to maternal nutritional status,* Journal of Nutrition (1994), 124(6S), 973S–980S.

Johnson, Allan A.; Knight, Enid M.; Edwards, Cecile H.; Oyemade, Ura Jean; Cole, O. Jackson; Westney, Ouida E.; Westney, Lennox S.; Laryea, Haziel; Jones, Sidney. *Selected lifestyle practices in urban African American women-relationships to pregnancy outcome, dietary intakes and anthropometric measurements,* Journal of Nutrition (1994), 124(6S), 963S–972S.

Edwards, Cecile H.; Johnson, Allan A.; Knight, Enid M.; Oyemade, Ura Jean; Cole, O. Jackson; Westney, Ouida E.; Jones, Sidney; Laryea, Haziel; Westney, Lennox S. *Pica in an urban environment,* Journal of Nutrition (1994), 124(6S), 954S–962S.

Knight, Enid M.; Spurlock, Bernice G.; Edwards, Cecile H.; Johnson, Allan A.; Oyemade, Ura Jean; Cole, O. Jackson; West, William L.; Manning, Malcolm; James, Hutchinson; et al. *Biochemical profile of African American women during three trimesters of pregnancy and at delivery,* Journal of Nutrition (1994), 124(6S), 943S–953S.

Johnson, Allan A.; Knight, Enid M.; Edwards, Cecile H.; Oyemade, Ura Jean; Cole, O. Jackson; Westney, Ouida E.; Westney, Lennox S.; Laryea, Haziel; Jones, Sidney. *Dietary intakes, anthropometric measurements and pregnancy outcomes,* Journal of Nutrition (1994), 124(6S), 936S–942S.

Edwards, Cecile H.; Knight, Enid M.; Johnson, Allan A.; Oyemade, Ura Jean; Cole, O. Jackson; Laryea, Haziel; Westney, Ouida E.; Westney, Lennox S. *Multiple factors as mediators of the reduced incidence of low birth weight in an urban clinic population,* Coll. Journal of Nutrition (1994), 124(6S), 927S–935S.

Edwards, Cecile H.; Knight, Enid M.; Johnson, Allan A.; Oyemade, Ura Jean; Cole, O. Jackson; Nolan, George; Westney, Ouida E.; West, William L.; Laryea, Haziel; et al. *Demographic profile, methodology, and biochemical correlates during the course of pregnancy,* Journal of Nutrition (1994), 124(6S), 917S–926S.

Nolan, George H.; Nahavandi, Masoud; Edwards, Cecile H.; Knight, Enid M.; Johnson, Allan A.; Oyemade, Ura Jean; Cole, O. Jackson; Westney, Ouida E.; Westney, Lennox S.; Winborne, Dvon. *Deoxyribonucleic acid, ribonucleic acid, and protein in the placentas of normal and selected complicated pregnancies,* Journal of Nutrition (1994), 124(6S), 1022S–1027S.

West, William L.; Knight, Enid M.; Edwards, Cecile H.; Manning, Malcolm; Spurlock, Bernice; James, Hutchinson; Johnson, Allan A.; Oyemade, Ura Jean; Cole, O. Jackson; et al. *Maternal low level lead and pregnancy outcomes,* Journal of Nutrition (1994), 124(6S), 981S–987S.

Jenkins, M. Young; Mitchell, G. Vaughan; Vanderveen, John E.; Adkins, J. S.; Edwards, C. H. *Effects of dietary protein and lecithin on plasma and liver lipids and plasma lipoproteins in rats,* Nutrition Reports International (1983), 28(3), 621–634.

Ganapathy, Seetha N.; Booker, Lovie K.; Craven, Richard; Edwards, Cecile H. *Trace minerals, amino acids, and plasma proteins in adult men fed wheat diets,* Journal of the American Dietetic Association (1981), 78(5), 490–497.

Edwards, Cecile H.; Wade, Wilda D.; Freeburne, Mary M.; Jones, Evelyn G.; Stacey, Robert E.; Sherman, Larry; Seo, Chung-Woon; Edwards, Gerald A. *Formation of methionine from α-amino-n-butyric acid and 5'-methylthioadenosine in the rat,* Journal of Nutrition (1977), 107(10), 1927–1936.

Obizoba, Ikemefuna Christopher; Adkins, James S.; Edwards, Cecile H. *Biological evaluation of the protein quality of wheat-soy-beef mixtures in weanling rats,* Sch. Hum. Ecol., Nutrition Reports International (1977), 15(6), 667–680.

Obizoba, Ikemefuna Christopher; Edwards, Cecile H.; Adkins, James S. *Biochemical evaluation of wheat-soy-beef mixtures fed weanling rats,* Nutrition Reports International (1977), 15(6), 659–666.

Edwards, Cecile H.; Rawalay, Surjan S.; Edwards, Gerald A. *Intermediary metabolism of methionine,* Journal of the Elisha Mitchell Scientific Society (1973), 89(3), 206–213.

Edwards, Cecile H.; Rawalay, Surjan S.; Higginbotham, Curtis; Edwards, Gerald A. *Distribution in rat tissues and urine of methionine labeled with carbon-14 in positions 1-,2-,3-,4-, or methyl carbon or with sulfur-35,* Journal of the Elisha Mitchell Scientific Society (1972), 88(4), 267–274.

Edwards, Cecile H.; Rawalay, Surjan S.; Edwards, Gerald A. *Formation of taurine-14C from methionine-methyl-14C,* Journal of the Elisha Mitchell Scientific Society (1971), 87(4), 237–238.

Edwards, G. A.; Jones, Evelyn G.; Higginbotham, C.; Edwards, Cecile H. *Method of correcting for the absorption of carbon-14 in animal tissues,* International Journal of Applied Radiation and Isotopes (1971), 22(5), 309–311.

Edwards, Cecile H.; Booker, Lovie K.; Rumph, Cordella H.; Craven, Richard; Ganapathy, Seetha N. *Utilization of wheat by adult man: excretion of vitamins and minerals,* American Journal of Clinical Nutrition (1971), 24(5), 547–555.

Edwards, Cecile H.; Booker, Lovie K.; Rumph, Cordella H.; Wright, Walter G.; Ganapathy, Seetha N. *Utilization of wheat by adult man: nitrogen metabolism, plasma amino acids and lipids,* American Journal of Clinical Nutrition (1971), 24(2), 181–193.

Edwards, Gerald A.; Rawalay, Surjan Singh; Edwards, Cecile H. *Formation of 5-methylthioadenosine and other sulfur-containing compounds from methionine-35S in the rat,* Proc. Int. Congr. Nutr., 7th (1967), Meeting Date 1966, 5, 158–159.

Edwards, Cecile Hoover; Gadsen, Evelyn L.; Higginbotham, Curtis; Edwards, Gerald A. *Utilization of methionine by the adult rat. X. Incorporation of methionine into tissue proteins,* Journal of the Elisha Mitchell Scientific Society (1966), 82(1), 12–18.

Edwards, Cecile H.; Gadsden, Evelyn L.; Edwards, Gerald A. *Utilization of methionine by the adult rat. II. Absorption and tissue uptake of L-and DL-methionine,* Journal of Nutrition (1963), 80(1), 69–74.

Edwards, Cecile Hoover; McDonald, Solona; Mitchell, Joseph R.; Jones, Lucy; Mason, Lois; Trigg, Louise. *Effect of clay and corn-starch intake on women and their infants,* Journal of the American Dietetic Association (1964), 44, 109–115.

Gadsden, Evelyn L.; Edwards, Cecile H.; Webb, Alfreda J.; Edwards, Gerald A. *Autoradiographic patterns of methionine-2-14C and methionine-methyl-14C in tissues of the adult rat,* Journal of Nutrition (1965), 87(2), 139–143.

Edwards, Cecile H.; Gadsden, Evelyn L.; Edwards, Gerald A. *Methionine and homocysteine as protective agents against irradiation damage,* Metabolism, Clinical and Experimental (1964), 13(4), 373–380.

Edwards, Cecile H.; Gadsden, Evelyn L.; Edwards, Gerald A. *Effects of irradiation on the tissue uptake of methionine-2-14C and methionine-methyl-14C,* Radiation Research (1964), 22 116–125.

Edwards, Cecile H.; Edwards, Gerald A.; Gadsden, Evelyn L. *Utilization of methionine by the adult rat. VII. The methyl carbon of methionine as a source of carbon in cholesterol,* Journal of the Elisha Mitchell Scientific Society (1963), 79, 108–111.

Edwards, Cecile H.; Edwards, Gerald A.; Gadsden, Evelyn L. *Tomatine and digitonin as precipitating agents in the estimation of cholesterol,* Anal. Chem. (1964), 36(2), 420–421.

Edwards, Cecile H.; Gadsden, Evelyn L.; Edwards, Gerald A. *Utilization of methionine by the adult rat. IV. Distribution of the methyl carbon of methionine in tissues, blood, expired carbon dioxide, and excreta,* Metabol. Clin. Exptl. (1963), 12, 951–958.

Edwards, Cecile H.; Gadsden, Evelyn L.; Edwards, Gerald A. *Utilization of methionine by the adult rat. III. Early incorporation of methionme-methy-14C and methionine-2-14C into rat tissues,* Journal of Nutrition (1963), 80(2), 211–216.

Edwards, Cecile H.; Rice, James O.; Jones, James; Seibles, Lawrence; Gadsden, Evelyn L.; Edwards, Gerald A. *Chromatography of compounds of biological interest on glass fiber, paraffin-coated, and untreated cellulose paper,* Journal of Chromatography (1963), 11(3), 349–354.

Edwards, Cecile H.; Gadsden, Evelyn L.; Edwards, Gerald A. *Utilization of methionine by the adult rat. I. Distribution of the alpha carbon of DL-methionine-2-C14 in tissues, tissue fractions, expired carbon dioxide, blood, and excreta,* Journal of Nutrition (1960), 72, 185–195.

Gadsden, Evelyn L.; Edwards, Cecile H.; Edwards, Gerald A. *Paper chromatography of certain vitamins in phenol and butanol-propionic acid-water solvents,* Anal. Chem. (1960), 32, 1415–1417.

Edwards, Cecile H.; Gadsden, Evelyn L.; Carter, Lolla P.; Edwards, Gerald A. *Paper chromatography of amino acids and other organic compounds in selected solvents,* Journal of Chromatography (1959), 2, 188–198.

Edwards, Cecile Hoover; Allen, Cordella Hill. *Cystine, tyrosine, and essential-amino-acid content of selected foods of plant and animal origin,* Journal of Agricultural and Food Chemistry (1958), 6, 219–223.

Edwards, Cecile H.; Carter, Lolla P.; Outland, Charlotte E. *Cystine, tyrosine, and essential-amino-acid contents of selected foods,* Journal of Agricultural and Food Chemistry (1955), 3, 952–957.

Betty Wright Harris

Harris, Betty W.; Singleton, Jannie L.; and Coburn, Michael D. *Picrylamino-substituted heterocycles. VI. Pyrimidines,* J. of Heterocyclic Chem. 10, 167–171 (April 1973).

Harris Betty W.; and Coburn, Michael. *Reaction of 2-aminopyridine with picryl halides."* J. Heterocyclic Chem. 13, 845–851 (August 1976).

Cady, H. H.; Coburn, M. D.; Harris, B. W.; and Rogers, R. N. *Synthesis and thermochemistry of ammonium-2,4,5-trinitroimidazole.* Los Alamos Scientific Laboratory Report, LA-6802-MS, (July 1977).

Harris, Betty W. *Carbon-13 NMR analyses of TATB and related compounds in sulfuric acid.* Los Alamos Scientific Laboratory Report, LA-7572-MS, (April 1979).

Harris, Betty W. *The chemistry of TATB and related compounds in sulfuric acid.* Los Alamos Scientific Laboratory Report LA-8629-MS (March 1981).

Harris, Betty W. *Sulfuric formation and deposition in combined cycle steam generation systems.* Solar Turbines Inc. Research Laboratories report SR-M-4769-00 (March 1983).

Harris, Betty W. and Johnson, T. L. *Soot deposition and growth in combined cycle systems.* Solar Turbines Inc. Research Laboratories report SR83-M-4821-00 (July 1983).

Harris, B. W. *Oil HE compatibility study—Selected safing fluids for damaged explosive assemblies.* Propellants, Explosives and Pyrotechnics 2, 9–11 (1984).

Harris, Betty W. *Oxidation of carbonaceous materials - RACER.* Solar Turbines, Inc. Research Laboratories report SR84-M-4909-00 (March 1984).

Harris, B. W. *Test results—Controlled condensation method of measuring SO2/SO3 concentrations in the exhaust of the racer subscale-hot-loop*. Solar Turbine, Inc. Research Laboratories report SR84-M-4909 (June 1984).

Harris, Betty W. *TATB—Strong basic reactions provide soluble derivatives for a simple, qualitative high explosive spot test*. J. Energetic Materials 3, 81–131 (1985).

Harris, Betty W. *Spot test for 1 3.5-triamino-2.4.6-trinitrobenzene. TATB*. U.S. Patent No. 46,18,452 (October 1986).

Coburn, Michael D.; Harris, Betty W.; Kien-Yin Lee; Stinecipher, Mary M.; and Hayden, Helen H. *Explosive synthesis at Los Alamos*. Ind. Eng. Chem. Prod. Res. Dev. 25, 68–72 (1986).

Harris, Betty W. *Conversion of sulfur dioxide to sulfur trioxide in gas turbines*, 2, J. Engineering for Gas Turbines and Power, 1990.

Harris, Betty W.; Archuleta, J. G.; King, W. F.; and Baytos, J. *Analysis of soil contaminated with explosives,* Los Alamos National Laboratory Report LA-11505-MS, (February 1989).

Mary Elliott Hill

Hill, Carl M.; Woodberry, Rudolph; Hill, Mary E.; Williams, Albert O. *Reduction with lithium aluminum hydride. II. Lithium aluminum hydride reduction of aryloxyalkylketene monomers and dimers,* Journal of the American Chemical Society (1959), 81, 3372–3374.

Hill, Carl M.; Woodberry, Rudolph; Simmons, Doris E.; Hill, Mary E.; Haynes, Lonnie. *Grignard reagents and unsaturated ethers. VII. The synthesis, properties, and reaction of β-substituted vinyl ethers with aliphatic and aromatic Grignard reagents,* Journal of the American Chemical Society (1958), 80, 4602–4604.

Hill, Carl M.; Haynes, Lonnie; Simmons, Doris E.; Hill, Mary E. *Grignard reagents and unsaturated ethers. VI. The cleavage of diallyl ethers by aliphatic and aromatic Grignard reagents,* Journal of the American Chemical Society (1958), 80, 3623–3625.

Hill, Carl M.; Simmons, Doris E.; Hill, Mary E. *Grignard reagents and unsaturated ethers. V. Mode of cleavage of α- and γ-substituted allyl ethers by Grignard reagents,* Journal of the American Chemical Society (1955), 77, 3889–3892.

Hill, Carl M.; Senter, Gilbert W.; Haynes, Lonnie; Hill, Mary E. *Grignard reagents and unsaturated ethers. III. Reaction of*

Grignard reagents with cyclic unsaturated ethers, Journal of the American Chemical Society (1954), 76, 4538–4539.

Hill, Carl M.; Hill, Mary E. *Low-pressure hydrogenation and several properties of methyl- and butylketene dimers,* Journal of the American Chemical Society (1953), 75, 4591.

Hill, Carl M.; Hill, Mary E. *Preparation and properties of pentamethyleneketene monomer and dimer,* Journal of the American Chemical Society (1953), 75, 2765–2766.

Hill, Carl M.; Haynes, Lonnie; Simmons, Doris E.; Hill, Mary E. *Grignard reagents and unsaturated ethers. II. Reaction of Grignard reagents with β,γ -unsaturated ethers,* Journal of the American Chemical Society (1953), 75, 5408–5409.

Hill, Carl M.; Hill, Mary E.; Williams, Albert O.; Shelton, Essie M. *The synthesis, properties, and catalytic hydrogenation of several aryloxy substituted ketene monomers and dimers,* Journal of the American Chemical Society (1953), 75, 1084–1086.

Hill, Carl M.; Haynes, Lonnie; Hill, Mary E. *The catalytic hydrogenation of alkylketene dimers,* Journal of the American Chemical Society (1952), 74, 3423.

Hill, Carl M.; Hill, Mary E.; Williams, Albert O.; Shelton, Essie M. *The synthesis, properties, and catalytic hydrogenation of several aryloxy substituted ketene monomers and dimers,* Journal of the American Chemical Society (1953), 75, 1084–1086.

Hill, Carl M.; Hill, Mary E.; Schofield, Helen I.; Haynes, Lonnie. *Ozonization and catalytic hydrogenation of cyclohexylketene dimer and (ω -cyclohexylalkyl) ketene dimers,* Journal of the American Chemical Society (1952), 74, 166–167.

Hill, Carl M.; Walker, Robert A.; Hill, Mary E. *The reaction of Grignard reagents withα,β -unsaturated ethers,* Journal of the American Chemical Society (1951), 73, 1663–1664.

Hill, Carl M.; Schofield, Helen I.; Spriggs, Alfred S.; Hill, Mary E. *The synthesis, properties, and dehydrohalogenation of some α -phenoxy and 2,4-dichlorophenoxy-substituted acid chlorides,* Journal of the American Chemical Society (1951), 73, 1660–1662.

Hill, Carl M.; Senter, Gilbert W.; Hill, Mary E. *Preparation, properties, and reactions of six chlorine-substituted phenoxyketene monomers,* Journal of the American Chemical Society (1950), 72, 2286–2287.

Lynda Marie Jordan

Jordan, Lynda. Chemistry Department, *Characterization of the human placental phospholipase A2*. Book of Abstracts, 219th ACS National Meeting. San Francisco, CA, March 2630, 2000, SOCED-008. American Chemical Society, Washington, DC.

Radvanyi, Francois; Jordan, Lynda; Russo-Marie, Francoise; Bon, Cassian. *A sensitive and continuous fluorometric assay for phospholipase A2 using pyrene-Iabeled phospholipids in the presence of serum albumin,* Analytical Biochemistry (1989), 177(1), 103–109.

Jordan, L. M.; Russo-Marie F. *Purification and partial characterization of phospholipase A2 isoforms from human placenta,* Journal of chromatography (1992), 597(1–2), 299–308.

Sinah Estelle Kelley

Nelson, George E. N.; Traufler, Donald H.; Kelley, Sinah E.; Lockwood, Lewis B. *Production of itaconic acid by aspergillus terreus in 20-liter fermentors,* Industrial & Engineering Chemistry (1952) 44 (5), 1166–1168.

Rhodes, R. A.; Moyer, A. J.; Smith, M. L.; Kelley, S. E. *Production of fumaric acid by Rhizopusarrhizus,* Appl Microbiol. (1959) Mar 7(2): 74–80.

Reatha Clark King

King, R. C. "Fluorine Flame Calorimetry," *Experimental Thermochemistry,* Chapter 15, III, S. Sunner, ed., International Union of Pure and Applied Chemistry, Division of Physical Chemistry, Commission on Thermodynamics and Thermochemistry, 1976, with G.T. Armstrong.

King, Reatha Clarke; Armstrong, George T. *Fluorine flame calorimetry. III. Heat of formation of chlorine trifluoride at 298.15.deg.K,* Journal of Research of the National Bureau of Standards, Section A: Physics and Chemistry (1970), 74(6), 769–779.

King, Reatha Clarke; Armstrong, George Thomson. *Constant pressure flame calorimetry with fluorine. II. Heat of formation of oxygen difluoride,* Journal of Research of the National Bureau of Standards, Section A: Physics and Chemistry (1968), 72(2), 113–131.

King, R. C.; Armstrong, G. T. *Heat of combustion and heat of formation of aluminum carbide,* Journal of Research of the National Bureau of Standards, Section A: Physics and Chemistry (1964), 68A(6), 661–668.

King, R. C.; Kleppa, O. J. *A thermochemical study of some selected Laves phases,* Acta Metallurgica (1964), 12(1), 87–97.

Kleppa, O. J.; King, R. C. *Heats of formation of the solid solutions of zinc, gallium, and germanium in copper,* Acta Metallurgica (1962), 10, 1183–1186.

King, R. C. *Fluorine flame calorimetry. III. The heat of formation of chlorine trifluoride*, J. Res. NBS 74A, 113 (1968), with G. T. Armstrong.

King, R. C.; Armstrong, G. T. Constant *Pressure flame calorimetry with fluorine. II. The heat of formation of oxygen difluoride,* J. Res. NBS 72A, 113 (1968).

King, R. C. *The heat of formation of aluminum carbide,* Technical News Bulletin, National Bureau of Standards, February 1965.

King R. C.; Armstrong, G. T. *Heat of combustion and heat of formation of aluminum carbide*, J. Res. NBS 68A, 661 (1964).

King, R. C.; Kleppa,O. J.: Hersh L. S. *Studies of fused salts. III. Heats of mixing silver nitrate mixtures*, J. Chem. Phys. 35, 1975 (1961).

Katheryn Emanuel Lawson

Lawson, Katheryn E. *Infrared absorption of inorganic substances,* 1961, Reinhold Publishing, New York.

Lawson, Katheryn E. *Optical studies of electronic transitions in hexa- and tetracoordinated* mn^{2+} *crystals,* J. Chem. Phys. 47, 3627 (1967).

Lawson, Katheryn E. *Electronic absorption spectra of* Mn^{2+}, Fe^{2+} *and* Co^{2+} *in dihydrated halide crystals,* J. Chem. Phys. 44, 4159 (1966).

Morosin, Bruno; Lawson, Katheryn. *The configuration and electronic absorption spectra of tetracloro- and tetrabromocuprate ions,*

Journal of Molecular Spectroscopy 12, January 1, 1964, 99–117.

Lawson, Katheryn L.; Kahn, Milton. *Adsorption and solvent extraction procedures for the separation of carrier-free indium from cadmium,* Journal of Inorganic and Nuclear Chemistry 1 issue 5 (1957), 87–92.

Jennie R. Patrick

Patrick, J. R., and Reid, R. C., *Superheat-limit temperature of polar liquids*, Ind. Eng. Chem. Fundamentals, November 1981.

Patrick, J. R, *Supercritical extraction technology,* NOBCChE Proceedings, 1981.

Patrick, J. R., and Palmer, F., *Supercritical extraction (SCE) of dixylenol sulfone (DXS),* Supercritical Fluid Technology (Elsevier), 1985.

D'Souza, Rupert; Patrick, Jennie R.; Teja, Amyn S. *High pressure phase equilibria in the carbon dioxide-n-hexadecane and carbon dioxide-water systems,* Canadian Journal of Chemical Engineering (1988), 66(2), 319–323.

Patrick, J. R. *Supercritical extraction of chlorinated biphenyls (PCB's) from transformer oil (l0-C),* Nov. 8, 1981, RD-13, 199.

Patrick, J. R. *Supercritical extraction of dichlorodimethylsilane from trichloromethylsilane,* Jan. 16, 1981, RD-13, 377.

Patrick, J. R. *Method of improving the quality of lexan-polycarbonate resin made from interfacial polymerization,* Nov. 23, 1981, RD-14, 050.

Patrick, J. R. *Purification of bisphenol A (BPA) via a supercritical fluid mixture,* Dec. 7, 1981, RD-14, 112.

Patrick, J. R. *Use of acetone for the purification of lexan-polycarbonate resin,*" RD-14, 231.

Johnnie Hines Watts Prothro

Watts, Johnnie H., and Griswold, Ruth M. *Enzyme and ascorbic acid contents of fresh and frozen pineapple.* Food Research 18:1~2-168 (1953).

Watts, Johnnie H., and Griswold, Ruth H. *Enzyme inactivation: Relation of rates of inactivation of peroxidase, catecholase, and ascorbase to oxidation of ascorbic acid in potatoes and parsnips.* Agriculture and Food Chemistry I: 569–574 (1953) J.

Watts. Johnnie H.; Booker, Lovie K.; Wilson, Janet V.; Williams, E. Geraldine; and Wright, Walter G. *Availability of energy and essential amino.acids in pork and peanut butter.* Federation Proceedings 16: 401 (1957).

Watts, Johnnie H., Booker, Lovie K.; Wright, Walter G.; and Williams, E. Geraldine. *The effect of various drying methbds on the nitrogen and fat contents of biological materials.* Food Research 21; 528–533 (1956).

McAfee, Janet Wilson, and Watts, Johnnie H. *Effects of non-antiscorbutic reducing substances upon the ascorbic acid content of baked potatoes.* Food -Research 23: 114–118 (1958).

Watts, Johnnie H., and Graham, Donald C.W. *Calcium excretion in human subjects at two levels of milk intake.* J. Dairy Science 41: 1224–1229 (1958).

Watts, Johnnie H. *Availability of amino acids from foods.* Nutrition News (October 1958).

Watts, Johnnie H.; Booker, Lovie K.; McAfee, Janet W.; Williams, E. Geraldine; Wright, Walter G.; and Jones, Fred, Jr. *Biological availability of essential amino acids to human subjects. I. Whole egg, pork muscle and peanut butter.* J. Nutrition 67: 483–496 (1959).

Watts, Johnnie H., Booker, Lovie K.; McAfee, Janet W.; Williams, E. Geraldine; Wright, Walter G.; and Jones Fred, Jr. *Biological availability of essential amino acids to human subjects. II. Whole egg, milk, and cottage cheese.* J. Nutrition 67; 497–508 (1959).

Watts, Johnnie H.; Allen, Cardella H.; and Booker, Lovie K. *Biological availability of essential amino acids to human subjects. III. Whole egg and beef muscle.* J. Am. Dietet. Assoc. 36: 45–54 (1960)

Watts, Johnnie H.; Swendseid, Marian E.; Harris, Cheryl S.; and Tuttle, Stewart G. *An evaluation of the FAO amino acid reference pattern in human nutrition. I. Studies with young men.* J. Nutrition 75: 295–302 (1961).

Watts, Johnnie H., Swendseid, Marian E.; Harris, Cheryl S.; and Tuttle, Stewart G. *Comparison of essential amino acid patterns.* Federation Proceedings 19:12 (1960).

Watts, Johnnie H., Graham, Donald C. W.; Jones, Fred, Jr.; Adams, F. JoAnn; and Thompson, D. J. *Fecal solids excreted by young men following the ingestion of dairy foods,* Am. J . Digestive Diseases 8: 364–375 (1963).

Tolbert, Bernadine, and Watts, Johnnie H. *The phenylalanine requirements of women consuming a minimal tyrosine diet and the sparing effect of tyrosine on the phenylalanine requirement.* J. Nutrition 80: 111–116 (1963).

Watts, Johnnie H. *Comparison of metabolic responses to the FAO and milk patterns of essential amino acids fed human subjects.* Abstract. Sixth International Congress of Nutrition, Edinburgh, August 9–15, 1963.

Watts, Johnnie H., Tolbert, Bernadine; and Ruff, Willie L. *Nitrogen balances for young adult males fed two sources of nonessential nitrogen at two levels of total nitrogen intake.* Metabolism l3: 172–180 (1964).

Watts, Johnnie H., Mann, Antoinette N.; Bradely, Lois; and Thompson, Daniel J. *Nitrogen balances of men over 65 fed the FAO and milk patterns of essential amino acids.* J. Geontlogy 19: 370–374 (1964).

Watts, Johnnie H., Tolbert, Bernadine; and Ruff, Willie L. *Nitrogen balances of young men fed selected amino acid patterns. I. FAO reference pattern, a modification of the FAO pattern, -and wheat flour pattern.* Can. J. Biochem. 42: 1437–1444 (1964).

Watts, Johnnie H. *Evaluation of protein in selected american diets.* J. Amer. Dietet. Assoc. 46: 116–119 (1965).

Watts, Johnnie H.; Bradley, Lois; and Mann, Antoinette N. *Total N, urea and ammonia excretions of human male subjects fed several nonessential amino acids singly as the chief source, of nonspecific N.* Metabolism 14: 504–515 (1965).

Prothro, Johnnie; MacKellar, Ingebore; Reyes, Nemesia; Linz, Maria; and Chou, Chuan. *Utilization of nitrogen, energy, and sulfur by adolescent boys fed three levels of protein.* J. Nutrition. l03: 786–791 (1973).

Pace, Ralphenia Diggs; Chen, Lie-wen Hsu; and Prothro, Johnnie. *Effect of dietary calcium and fat on calcium and lipid content of rat tissues.* Nutrition Reports International-7: 121–132 (1973).

Prothro, Johnnie W.; Mickles, Mary; and Talbert, Bernadine. *Nutritional status of a population sample in Macon County.* Am . J. Clin. Nutr. 29: 94–104 (1976).

Gladys W. Royal

Royal, G. W.; Hunter, G. W.; and White, B. T. *A division directed approach to introductory chemistry at the secondary school level.* School Science and Mathematics, 62 (1962): 269–275. doi: 10.1111/j.1949-8594.1962.tb08704.x.

Royal, Gladys Williams. *The influence of rations containing sodium acetate and sodium propionate on the composition of tissues from feeder lambs.* Thesis (PhD), Ohio State University, 1954.

Cheryl L. Shavers

Shavers, Cheryl L. *Same sky, different horizon: The role of two-year colleges in career success,* Journal of Chemical Education (1999), 76(4), 458. Division of Chemical Education of the American Chemical Society,

Shavers, Cheryl Lynn. *Theoretical and experimental studies in crystal chemistry,* Arizona State Univ., Tempe, AZ, USA. Avail. Univ. Microfilms Int., Order No. 8115607. (1981), 228 pp. From: Diss. Abstr. Int. B 1981, 42(2), 650.

O'Keeffe, M.; Shavers, C. L.; Hyde, B. G. *Cation packing and molar volume in oxides and nitrides with the wurtzite structure,* Journal of Solid State Chemistry (1981), 39(2), 265–267.

Shavers, C. L.; Parsons, M. L.; Deming, S. N. *Simplex optimization of chemical systems,* Journal of Chemical Education (1979), 56(5), 307–309.

Linda Meade Tollin

Tropp, B. E.; Meade, L. C.; and Thomas, P. J. *Consequences of expression of the "relaxed" genotype of the RC gene -lipid synthesis.* Journal of Biological Chemistry 245: 855–859, 1970.

Meade, LC *Biochemical Inactivity*. Journal of Interdisciplinary Studies (CUNY) 1:71-72,1974.

Nunn, W.D.; Tropp, B. E.; and Meade, L. C. *Lipid synthesis in stringent escherichia coli: an artifact m acetate labeling ofphospholipids during a shift down in growth rate*. Journal of Bacteriology 121: 396–399 (1975).

Meade-Tollin, L. C., Pipes, B.; Anderson, S. J.; Seftor, E.; and Hendrix, M. J. C. *A comparison of levels of intrinsic single-strand breaks / alkali-labile sites in invasive human melanoma cells*. Cancer Letters, 53, 45–54 (1990).

Tsang, T. C.; Chu, Y. W.; Powell, M. B.; Kittelson, J.; Meade-Tollin, L.; Hendrix, M. T.; Bowden, G. T. *v-jun oncogene suppresses both phorbol ester-induced cell invasion and strome lysin gene expression in a mouse papilloma cell line*. Cancer Research, 54(4), 882–886 (1994).

Meade-Tollin, L. C.; Boukamp, P.; Fusenig, N. E.; Bowen, C. P. R.; Tsang, T. C.; Bowden, G. T. *Differential expression of matrix metalloproteinases in activated c-rasHa-transfected immortalized human keratinocytes*, British Journal of Cancer, 77(5), 724–730 (1998).

Meade-Tollin, L. C.; Way, D.; and Witte, M. *Expression of multiple matrix metalloproteases and urokinase type plasminogen activator in cultured Kaposi sarcoma cells*. Acta histochemica, 101, 305–316 (1999).

Van Noorden, C. J. F.; Jonges, T. G. N.; Meade-Tollin, L. C.; Smith, R. E.; and Koehler, A. In vivo *inhibition of cysteine proteinases delays the onset of growth of human pancreatic cancer explants*. British Journal of Cancer, 82(4), 931–936 (2000).

Tsang, T. C.; Brailey, J. L.; Vasanwala, F. H.; Wu, R. S.; Liu, F.; Clark, P. R.; Luznick, L.; Stopeck, A. T.; Mersh, E.; Akporiaye, E. T.; Meade-Tollin, L. C.; and Harris, D. T. *Construction of new amplifier expression vectors for high levels ofIL-2 gene expression*. International Journal of Molecular Medicine 5 (3), 295–300 (2000).

Meade-Tollin, L. *Meeting report, biological basis of antiangiogenic therapy*, Acta histochemica 102, 1–11 (2000).

Meade-Tollin, L. C., and Van Noorden, C. J. F. *Time-lapse phase contrast video microscopy of directed migration of human microvascular endothelial cells on matrigel.* Acta histochemica 102, 299–307 (2000).

McEarchern, J. A.; Kobie, J. J.; Mack, V.; Wu, R. S.; Meade-Tollin, L.; Arteaga, C. L.; Dumont, N.; Besselsen, D.; Seftor, E.; Hendrix, M. J. C.; Katsanis, E.; Akporiaye, E. T. *Invasion and metastasis of a mammary tumor involves TGF-p signaling.* International Journal of Cancer, 91 (1), 76–82 (2001).

Meade-Tollin, L. C.; Wijeratne, E.M. K.; Cooper, D.; Guild, M.; Jon, E.; Fritz, A.; Zhou, G.-X.; Whitesell, L.; Liang, J.-Y.; Leslie Gunatilaka, A. A. *Ponicidin and oridonin are responsible for the antiangiogenic activity of Rabdosia rubescens, a constituent of the herbal supplement, PC SPES.* Journal of Natural Products, 67(1): 2–4 (2004).

Meade-Tollin, L., and Martinez, J. D. *Loss of p53 and overexpression of EphA2 predict poor prognosis for ovarian cancer patients.* Cancer Biology and Therapy 6 (2) 288 (2007).

Peer-Reviewed Electronic Publications

Video Clip of Time-lapse phase contrast video microscopy of directed migration of human

microvascular endothelial cells on matrigel. Acta histochemica web site. http://www.urbanfischer.de/journals/actahist. (Follow link for Video Preview, 3/2000.)

AVI video clip of endothelial cell migration on web site for UA department of Toxicology,

SWEHSC, Experimental Pathology Core, at this address: http://swehsc.pharmacy.arizona.eduJexppath/core/equip/compix_examples.htm 1

Inventions

Co-inventor on U.S. Provisional Application #501711,249, "Stem Cell Fusion Model of Carcinogenesis," 2006.

Invited Scholarly Publications and Presentations

Meade-Tollin, L. 6th Annual Cancer Drug Symposium, Arizona Cancer Center. April 2004. Tucson, AZ.

Meade-Tollin, L. Visualization of Angiogenesis, AACR Special Conference, "Proteases, Extracellular Matrix, and Cancer," October, 2002, Hilton Head, SC.

Meade-Tollin, L. Visualization of Angiogenesis in vitro. The 6th Joint Meeting of the Japanese Society of Histochemistry and Cytochemistry and the Histochemical Society. July 18–21, 2002, University of Washington, Seattle, WA.

Meade-Tollin, L. Science, Cancer, and Philosophy, University of Arizona Africana Studies Department, April 2002.

Meade-Tollin, L. University of Arizona, African Americans in Life Sciences, 1st Annual Forum, Black Professionals Blazing the Path for Success, November 2001.

Meade-Tollin, L. Telik Corp., San Francisco, CA, March 2001.

Van Noorden, C. J.F.; Meade-Tollin, L. C.; Bosman. F. T. Metastasis, American Scientist, 86, #2,130, 1998.

Meade-Tollin, L. Rapid staining of gelatin zymograms with GelCode Blue Stain Reagent. Pierce Previews,2 (2) 2–5, 1998.

Meade-Tollin, L. Henry Hill Award Lecture: Matrix Metalloproteinases: A Link between Tumor Cell Metastasis and Angiogenesis? Proceedings of the 25th Annual Meeting of NOBCChE, Dallas TX, 1998.

NHLBI Cardiovascular Minority Research Supplement Awardee Session, Anaheim, CA, 1111995.

II. Morehouse School of Medicine, Biochemistry Department, Atlanta, GA, August 1986.

Atlanta University Department of Chemistry, Atlanta, GA, 1978.

National Black Chemists and Chemical Engineers, 5th Annual Meeting, Boston, MA, 1978.

Texts

Meade, L. C., contributor. *What People Eat*, Vol. 1, Preliminary Edition, Harvard University Press, 1974.

Rubye Prigmore Torrey

Hill, Carl M.; Prigmore, Ruby M.; Moore, George J. *Grignard reagents and unsaturated ethers. IV. The synthesis and reaction of*

several vinyl ethers with Grignard reagents, Journal of the American Chemical Society (1955), 77, 352–354.
Note: For other publications of Dr. Torrey, please consult reference databanks like the American Chemical Society's SciFinder.

HISTORICAL TIME LINE

1852 Philadelphia Quakers establish the Institute for Colored Youth as the first coeducational classical high school for African Americans. Myrtilla Miner, a white educator, founds the Normal School for Colored Girls in Washington, D.C.

Josephine Silone (Yates) is born in Mattick, New York.

1860 The black population of the United States totals 44,41,830—of which 488,070 are free and 3,953,760 are slaves. There are 2,225,086 women, 1,971,135 of whom are enslaved.

1863 President Lincoln issues the Emancipation Proclamation, liberating slaves in the Confederacy. The decree does not apply to the border states, and it exempts certain areas of Louisiana, Virginia, and West Virginia.

1865 Congress establishes the Freedmen's Bureau to coordinate aid and relief efforts, including education, for newly emancipated slaves. Two black colleges are founded: Atlanta University (Atlanta, Georgia) and the Shaw Institute (Raleigh, North Carolina).

1865 On December 6, the Thirteenth Amendment is ratified, ending slavery throughout the United States.

1866 Three black colleges are founded: Fisk University (Nashville, Tennessee), Rust College (Holly Springs, Mississippi), and Lincoln University (Jefferson City, Missouri).

1868 With financial support from Northern philanthropists, Hampton Institute is founded in Virginia to provide industrial education to former slaves to help them achieve self-sufficiency.

1868 States ratify the Fourteenth Amendment, granting citizenship to freed people and guaranteeing due process and equal protection of the law for all Americans, without regard to race or previous condition of servitude.

1869 Clark College (Atlanta, Georgia), Claflin College (Orangeburg, South Carolina), Straight College (now Dillard, New Orleans, Louisiana), and Tougaloo College (Mississippi) are founded.

1870 The black population of the United States is 4,880,009 or 12.7 percent of the total population. There are 2,486,746 women.

1871 Fanny Jackson (Coppin) becomes principal of the Institute for Colored Youth in Philadelphia. She is the first black woman to head an institution of higher learning in the United States.

1872 Alcorn A & M College (Lorman, Mississippi) is founded.
Beebe Steven (Lynk) is born on October 24 in Mason, TN.

1873 Bennett College (Greensboro, North Carolina), Wiley College (Marshall, Texas), and Alabama State College (Montgomery) are founded.

1875 Alabama A & M College (Normal), Knoxville College (Tennessee), and Lane College (Jackson, Tennessee) are founded.

1876 Prairie View A & M College (Texas) is founded.

1877 Jackson State College (Mississippi) is founded.

1879 Livingstone College (Salisbury, North Carolina) is founded.
Josephine Silone (Yates) graduates from Rhode Island Normal.
Josephine Silone (Yates) becomes professor and head of the Natural Sciences Department at Lincoln University (Jefferson City, Missouri), earning $1,000 per year.

1880 The African American population of the United States is 65,80,793 or 13.1 percent of the total population. Women total 3,327,678.

1880 Southern University (New Orleans, later Baton Rouge, Louisiana) is founded.

1881 Booker T. Washington founds Tuskegee Institute, based on the model of Hampton Institute, to provide moral and industrial training for black youth and to train teachers for the public schools.

1882 Virginia State College (Petersburg) is founded.

1886 Kentucky State College (Frankfort) is founded.

1887 Florida A & M College (Tallahassee) and Central State College (Wilberforce, Ohio) are founded.

1889 **Josephine Silone (Yates)** marries William Ward Yates.

1890 The black population of the United States is 7,488,676 or 11.9 percent of the total population. Women total 3.753,073.

1890 Savannah State College (Georgia) is founded.

1891 Delaware State College (Dover), North Carolina A & T College (Greensboro), and West Virginia State College (Institute) are founded.

1892 **Beebe Steven (Lynk)** graduates from Lane College.
Alice Augusta Ball is born on July 24.

1893 Meharry Medical College, founded in 1876 in Nashville, Tennessee, awards its first medical degrees to women: Georgianna Patten and Anna D. Gregg.
Beebe Steven marries Miles Lynk on April 13.

1896 In the landmark *Plessy v. Ferguson* case, the U.S. Supreme Court upholds a Louisiana law providing for separate railroad cars for blacks and whites. This decision establishes the legal basis for racial segregation in public facilities as long as "separate but equal" facilities are provided; the ruling becomes the legal foundation for Jim Crow segregation throughout the nation.
Eslanda Goode (Robeson) is born on December 15.

1896 South Carolina State College (Orangeburg) is founded.

1897 Spelman Seminary begins a College Department, with collegiate courses offered on Atlanta Baptist (Morehouse) campus.

1900 The black population of the United States is 8,833,994 or 11.6 percent of the total population. Women total 4,447,447.

1901 Grambling College (Louisiana) is founded.
Beebe Steven Lynk earns a PhC in pharmaceutical chemistry from the University of West Tennessee.

1905 **Angie Turner (King)** is born on December 9 in Elkhorn, West Virginia.

1907 **Mary Elliott (Hill)** is born on January 5 in South Mills, South Carolina.

1909 Tennessee State A & I University (Nashville) is founded.

1910 Black population of the United States is 9,827,763 or 10.7 percent of the total population. There are 4,941,882 women

1912 **Alice Ball** obtains a BS in pharmaceutical chemistry from the University of Washington.
Josephine Silone Yates dies.

1914 **Alice Ball** obtains a BS in pharmacy for the University of Washington.

1915 **Alice Ball** obtains an MS from the University of Hawaii.

1916 First African American to earn a doctorate in chemistry: St. Elmo Brady, University of Illinois.
Alice Ball dies.

1918 Elmer Samuel Imes awarded the doctorate in physics at the University of Michigan.

1920 **Eslanda Goode** obtains a BS in analytical chemistry from Columbia University.

1921 **Eslanda Goode** marries Paul Robeson on August 17.
Marie Maynard Daly is born on April 16.

1922 **Johnnie Hines Watts (Prothro)** is born on February 28 in Atlanta, Georgia.

1924 Spelman Seminary becomes Spelman College, for the first time offering college courses on its own campus.

1925 Sister Katherine Drexel and the Sisters of the Blessed Sacrament establish Xavier University in New Orleans; it is the only historically black Catholic college in the United States.

1925 **Mary Elliott (Hill)** marries Carl McClellan Hill.

1926 Carter G. Woodson organizes the first Negro History Week, which later becomes Black History Month.

1926 Bennett College, founded as a coeducational institution in 1873, becomes a college for women.

1926 National Technical Association founded.
Rubye Prigmore (Torrey) is born on February 18 in Sweetwater, Tennessee.
Gladys Williams (Royal) is born on August 26 in Dallas, Texas.
Esther A. Harrison (Hopkins) is born on September 16, in Stamford, Connecticut.
Cecile Hoover (Edwards) is born on October 26 in East St. Louis, Illinois.

1926 **Katheryn Emanuel (Lawson)** is born on September 15 in Shreevport LA.

1927 **Angie Turner (King)** obtains a BS in chemistry from West Virgina State University.

1929 **Mary Elliott (Hill)** obtains a BS in chemisty from Virginia State University.

1930 Black population of the United States is11,891,143 or 9.7 percent of the total population. There are 6,035,474 women.

1931 Percy Lavon Julian awarded the doctorate in chemistry at the University of Vienna.
Angie Turner (King) obtains a MS in Chemistry from Cornell University.

1932 Samuel Milton Nabrit, first African American awarded a doctorate in biological sciences (biology) at Brown University.

1932 Vernon Alexander Wilkerson, first African American awarded a doctorate (biochemistry) at the University of Minnesota.

1933 **Allene Johnson** is born on April 20 in Supply, North Carolina.

1934 **Jeannette Brown** is born on May 13 in New York, New York.
Carole Rodez aka Mary Antoinette (Schiesler) is born on December 13 in Chicago, Illinois.

1938 **Sinah Kelley** obtains a BA in chemistry from Radcliffe College.
Reatha Clark (King) is born on April 11 in Parvo, Georgia.

Gloria Long (Anderson) is born on November 5 in Althiemer, Arkansas.

1940 Black population of the United States is 12,865,578 or 10.8 percent of the total U.S. population. There are 6.596.480 women; 60 percent of all black women in the labor force are employed in domestic service, 10.5 percent are in other service work; only 1.4 percent work in clerical and sales positions, while 4.3 percent are in professional positions.

1940 Black scientists work on classified wartime research such as the Manhattan Project.

Betty Wright (Harris) is born on July 29 in northeastern Louisiana.

1941 **Johnnie Hinds Watts (Prothro)** obtains a BS in home economics from Spelman College.

Gladys Williams (Royal) obtains a BS in chemistry from Dillard University.

1942 **Marie Maynard Daly** obtains a BS in chemistry from Queens College.

1943 **Marie Maynard Daly** obtains a MS in chemistry from New York University.

Margaret Ellen Mayo (Tolbert) is born November 24 in Sulffolk, Virginia.

1944 **Rubye Prigmore (Torrey)** obtains a BS in chemistry from Tennessee State University.

Linda Meade (Tollin) is born August 16 in London, West Virginia.

1945 **Kathryn Emanuel (Lawson)** obtains a BS in chemistry from Dillard University.

Lilia Abron is born on March 8 in Memphis, Tennessee.

1946 **Angie Turner** marries Robert Elemore King on June 9.

Johnnie Hinds Watts (Prothro) obtains an MS in chemistry from Columbia University.

Cecile Hoover (Edwards) obtains a BS in chemistry from the Tuskeegee Institute.

1947 Receiving a degree from Columbia University, **Marie M. Daly** becomes the first African American woman to earn a PhD in chemistry.

Cecile Hoover (Edwards) obtains an MS in chemistry from the Tuskeegee Institute.

1947 **Sinah E. Kelley**, chemist, aids in mass production of penicillin.

Gladys Williams marries George C. Royal.

Katheryn Emanuel (Lawson) obtains an MS in chemistry from the Tuskeegee Institute.

Esther A. H. (Hopkins) obtains a BS in chemistry from Boston University.

1948 **Esther A. H. (Hopkins)** obtains an MS in chemistry from Howard University.

1949 **Johnnie Hinds Watts** marries Charles E. Prothro.

Jennie Patrick is born on January 1 in Gadsden, Alabama.

1950 Black population of the United States is 15,042,386 or 10 percent of the total population. There are 7,743,564 women.

1950 Forty-two percent of all black women in the labor force are employed in domestic service and 19.1 percent are in other service work; only 5.4 percent are in clerical and sales positions, and 5.7 percent are in professional positions.

Cecile Hoover (Edwards) obtains a PhD in chemistry from Iowa State University.

1951 **Cecile Hoover** marries Dr. Gerald A. Edwards in June.

1951 **Marie M. Daly**, chemist, works at the Rockefeller Institute for Medical Research.

1952 **Johnnie Hinds Watts Prothro** recevies a PhD in home economics from the University of Chicago.

1953 **Cheryl L. Shavers** is born in Texas.

1954 The Supreme Court issues its decision in *Brown v. Board of Education*, abolishing segregation and ordering that the states proceed "with all deliberate speed" to desegregate public schools; this decision overturns the "separate but equal" doctrine established in 1896 in *Plessy v. Ferguson*.

Katheryn Emanuel marries Kenneth Lawson.

Allene Johnson obtains a BS in chemistry from North Carolina College.

1955 **Angie Turner King** obtains a PhD in mathematics from the University of Pittsburg.
Rubye Prigmore marries Claude A. Torrey in September.

1956 **Jeannette Brown** obtains a BS in chemistry from Hunter College.
Lynda Marie Jordan is born on September 20 in Roxbury, Massachusetts.

1957 Congress passes the Civil Rights Act of 1957, the first such legislation since Reconstruction, giving the attorney general greater authority to handle interference with school desegregation; the Civil Rights Act also establishes a new Civil Rights Commission and provides that suits regarding black disfranchisement be heard in federal courts instead of state courts.
Gladys W. Royal obtains a PhD in chemistry from Ohio State University.
Katheryn Emanuel Lawson obtains a PhD in chemistry from the University of New Mexico.

1958 National Urban League announces initiative to encourage study of science among African Americans: establishes "Tomorrow's Scientists and Technicians" program.
Jeannette Brown obtains an MS in chemistry from the University of Minnesota. She is the first African American woman to obtain a degree in chemistry at the University of Minnesota.
Reatha Clark (King) obtains a BS in chemistry from Clark College.
Gloria Long (Anderson) obtains a BS in chemistry from Arkansas A&M Normal.

1959 **Esther A. Harrison** marries Ewell Hopkins on January 20.

1960 President Eisenhower signs the Civil Rights Act of 1960, prohibiting the intimidation of black voters and authorizing judges to appoint referees to oversee black voter registration.

1960 Black population of the United States is 18,871,831 or 10.5 percent of the total population. There are 9,758,423 women.

1960 Of all black women in the labor force, 32.5 percent are employed in domestic service, 21.4 percent are in other service positions, 10.8 percent are in clerical and sales, and 6 percent are in professional positions.

1960 Women's League of Science and Medicine, New York City, founded by Reva Green and Sarah Blow.
Reatha Clark (King) obtains an MS in chemistry from the University of Chicago.
Gloria Long marries Lenard Sinclar Anderson.
Betty Wright marries Alloyd A. Harris on July 8.

1961 **Marie Maynard Daly** marries Vincent Clark.
Reatha Clark marries N. Judge King.
Gloria Long Anderson obtains an MS in chemistry from Atlanta University.
Betty Wright Harris obtains a BS in chemistry from Southern University, Baton Rouge, Louisiana.

1963 **Reatha Clark King** obtains a PhD in chemistry from the University of Chicago.
Betty Wright Harris obtains an MS in chemistry from Atlanta University.

1964 President Lyndon Johnson signs the Civil Rights Act of 1964, increasing the authority of the attorney general to protect citizens against discrimination; the legislation denies federal funds to programs that discriminate and establishes the Equal Employment Opportunity Commission.
Linda Meade (Tollin) obtains a BS in chemistry from West Virginia State College.

1965 **Eslanda Robeson** dies on December 13.

1966 **Lilia Abron** obtains a BS in chemistry from LeMoyne College.

1967 **Esther A. H. Hopkins** obtains a PhD in chemistry from Yale University.
Margaret Ellen Mayo (Tolbert) obtains a BS in chemistry from Field Tuskegee University.

1968 **Rubye Prigmore Torrey** obtains a PhD in chemistry from Syracuse University. She is the first black woman to obtain the degree at Syracuse University
Gloria Long Anderson obtains a PhD in organo fluorene chemistry from the University of Chicago.

Margaret Ellen Mayo (Tolbert) obtains an MS in chemistry from Wayne State University.
Lilia Abron obtains an MS in environmental engineering from Washington University.

1968 **Gloria Long Anderson** obtains a PhD in chemistry from the University of Chicago.

1969 National Black Science Students Organization founded at the University of California, Los Angeles.
Mary Elliott Hill dies on February 16.
Mary Antoinette (Schiesler) obtains an MS in chemistry from the University of Tennesee, Knoxville.
Linda Meade (Tollin) obtains an MS in biochemistry from Hunter College.

1971 Smithsonian Institution sponsors an exhibit on African American scientists at the Anacostia Neighborhood Museum, Washington, D.C.

1972 Ad Hoc committee for the Professional Advancement of Black Chemists and Chemical Engineers conducts a survey of black professionals to ascertain interest in establishing "a formal organization dedicated to the professional advancement of Black chemists and chemical engineers."
Margaret Ellen Mayo marries Dr. Henry Hudson Tolbert.
Linda Meade (Tollin) obtains a PhD in biochemistry from Hunter College.
Lilia Abron obtains a PhD in chemical engineering from the University of Iowa.

1973 **Mary Antoinette Rodez** marries Alan Schiesler on October 20.
Jennie Patrick obtains a BS in chemical engineering from the University of California, Berkeley.

1974 First meeting the National Organization for the Professional Advancement of Black Chemists and Chemical Engineers (NOBCChE): first president is William Guillory.
Margaret Ellen Mayo Tolbert obtains a PhD in biochemistry from Brown University.

1975 NOBCChE establishes the Percy L. Julian Outstanding Research Award.

Betty Wright Harris obtains a PhD in analytical chemistry from the University of New Mexico.

1976 Henry Hill becomes the first black president of the American Chemical Society.

Cheryl Shavers obtains a BS in chemistry from Arizona State University.

1976 National Consortium for Graduate Degrees for Minorities in Science and Engineering is founded.

1977 **Esther A. H. Hopkins** obtains a JD from Suffolk College.

1978 National Network of Minority Women in Science is founded.

Linda Meade marries Dr. Gordon Tollin.

Lynda Marie Jordan obtains a BS in chemistry from North Carolina A & T University.

1979 **Jennie Patrick** is the first black woman in the U.S. to earn a PhD in chemical engineering (Massachusetts Institute of Technology.

1980 **Lynda Marie Jordan** obtains an MS in chemistry from Atlanta University.

1981 **Cheryl Shavers** obtains a PhD in solid state chemistry from Arizona State University.

1982 **Sinah Kelley** dies on December 21.

1985 **Lynda Marie Jordan** obtains a PhD in organic chemistry from MIT.

1987 Mae Jemison joins NASA, becoming the first African American woman astronaut.

1990 The Quality Education for Minorities Project releases a report finding that minority students are taught in "separate and decidedly unequal" schools; the report makes recommendations aimed at making schools more responsive to the needs of minority students.

1992 Mae Jemison becomes the first African American woman in space, traveling as the science mission specialist in the seven-member crew aboard the space shuttle *Endeavour*.

1993 President Clinton appoints a record number of African Americans to positions in his administration, including secretaries of agriculture, energy, and commerce; head of veterans affairs; and surgeon general.

1996 **Mary Antoinette Schiesler** dies.
2002 **Gladys W. Royal** dies on November 9.
2003 **Marie Maynard Daly** dies on October 23.
2004 **Angie Turner King** dies on February 28.
2005 **Cecile Hoover Edwards** dies on September 17.
2008 **Katheryn Emanuel Lawson** dies on September 25.
2009 Lisa P. Jackson is confirmed as chief administrator of the federal Environmental Protection Agency, becoming the first African American to hold that office.
Johnnie Hinds Watts Prothro dies on June 6.

BIBLIOGRAPHY

African American Women Scientists

Davis, Marianna W. 1982. *Contributions of Black Women to America*, vol. II, Columbia SC: Kenday Press.

Gates, Henry Louis, and Evelyn Brooks Higginbotham. 2008. *The African American National Biography*. New York: Oxford University Press.

Jordan, Diann. 2006. *Sisters in Science: Conversations with Black Women Scientists about Race, Gender, and Their Passion for Science*. West Lafayette, IN: Purdue University Press.

Kessler, James H. 1996. *Distinguished African American Scientists of the 20th Century*. Phoenix, AZ: Oryx Press.

Krapp, Kristine. 1999. *Notable Black American Scientists*. Detroit: Gale.

Sammons, Vivian Ovelton. 1989. *Blacks in Science and Medicine*. New York: Hemisphere.

Schiesler, Toni. 1994. "My Mother's Power Was in Her Voice," in *I've Known Rivers–Lives of Loss and Liberation*, Sara Lawrence-Lightfoot. Reading, MA: Addison-Wesley Publishing, 197–223.

Smith, Jessie Carney, and Phelps, Shirelle, Eds. 1992. *Notable Black American Women*. Detroit: Gale.

Warren, Wini. 1999. *Black Women Scientists in the United States*. Bloomington: Indiana University Press.

Warren,Wini Mary Edwina. *Hearts and Minds Black Women Scientists in the United States 1900–1960*. PhD dissertation, Department of the History and Philosophy of Science (Indiana University, 1997), available on University Microfilm.

History

Manning, Kenneth R. 2007. "African Americans in Science," in, *Ideology, Identity, and Assumptions*, eds. Howard Dodson and Colin A. Palmer. New York: New York Public Library, 49–95.
Manning, Kenneth R. 1991. "The Complexion of Science," *Technology Review* (November-December): 61–69.
Pearson,Willie. 1985. *Black Scientists, White Society, and Colorless Science: A Study of Universalism in American Science*. Millwood, NY: Associated Faculty Press.
Pearson, Willie. 2005. *Beyond Small Numbers: Voices of African American PhD Chemists*. Amsterdam: Elsevier JAI.
Taylor, Julius H., and Morgan State College. 1955. *The Negro in Science*. Baltimore: Morgan State College Press.
Young, Herman A., and Young, Barbara H. 1974. *Scientists in the Black Perspective*. s.l: s.n.

Women Chemists

Ambros, Susan A., and Dunkle, Kristin L., Kazarus, Barbara B., Nair, Indira, and Harkus, Deborah A. 1997. *Journeys of Women in Science and Engineering*. Philadelphia, Temple University Press.
Bailey, Martha J. 1998. *American Women in Science: 1950 to the Present: A Biographical Dictionary*. Santa Barbara, CA: ABC-CLIO.
Fort, Deborah C., Bird, Stephanie J., and Didion, Catherine Jay, eds. 1993. *A Hand Up: Women Mentoring Women in Science*. Washington, D.C.: Association for Women in Science.
Grinstein, Louise S., Rose K. Rose, and Miriam H. Rafailovich. 1993. *Women in Chemistry and Physics: A Biobibliographic Sourcebook*. Westport, CT: Greenwood Press.
Herman, Christine. 2008. *Iota Sigma Pi: National Honor Society of Women in Chemistry*. Radford, VA: Iota Sigma Pi.

Hinkle, Amber S., and Jody A. Kocsis. 2005. *Successful Women in Chemistry: Corporate America's Contribution to Science.* Washington, DC: American Chemical Society.

Moore, Patricia, Judith Love Cohen, and David A. Katz. 2005. *You Can Be a Woman Chemist.* Marina del Rey, CA: Cascade Pass.

Rayner-Canham, Marelene F., and Geoffrey William Rayner-Canham. 1998. *Women in Chemistry: Their Changing Roles from Alchemical Times to the Mid-Twentieth Century.* Washington, DC: American Chemical Society.

Sammons, Vivian Ovelton. 1990. *Blacks in Science and Medicine.* New York: Hemisphere Publishing.

Educators' Resources

Hanson, Sandra L. 2009. *Swimming against the Tide: African American Girls and Science Education.* Philadelphia: Temple University Press.

Jemison, Mae. 2001. *Find Where the Wind Goes: Moments from My Life.* New York: Scholastic.

Kahn, Jetty. 2000. *Women in Chemistry Careers.* Mankato, MN: Capstone Books.

Madden, Annette. 2000. *In Her Footsteps: 101 Remarkable Black Women.* New York: Gramacy Books.

Otha, Richard Sullivan. 2002. *Black Stars: African American Women Scientists and Inventors.* New York: John Wiley & Sons,.

St. John, Jetty. 1996. *African-American Scientists.* Mankato: Capstone.

Verheyden-Hilliard, Mary Ellen. 1985. *Scientist and Administrator, Antoinette Rodez Schiesler.* Bethesda, MD: The Equity Institute.

Verheyden-Hilliard, Mary Ellen. 1985. *Mathematician and Administrator, Shirley Mathis McBay.* Bethesda, MD: The Equity Institute.

Warmager, Paul, and Heltzel, Carl. 2007. "Alice A. Augusta Ball: Young Chemist Gave Hope to Millions," *ChemMatters* (February): 16.

The Double Bind: The Price of Being a Woman and Minority in Science http://www.eric.ed.gov/PDFS/ED130851.pdf (accessed March 10, 2011).

Science Activity Books for Educators

Bernstein, Leonard, Alan Winkler, and Linda Zierdt-Warshaw. 1998. *African and African American Women of Science*. Maywood, NJ: Peoples Pub. Group.

Warren, Rebecca Lowe, and Mary H. Thompson. 1995. *The Scientist Within You*. Eugene, OR: ACI Pub.

Warren, Rebecca Lowe, and Mary H. Thompson. 1996. *The Scientist Within You,* [Vol. 2]: *Women Scientists from Seven Continents : Biographies and Activities*. Eugene, Or: ACI Pub.

Media

Archives of Women in Science and Engineering (Iowa State University). 2001. *The Women in Chemistry Oral History Project*. Ames : Iowa State University. http://lib.iastate.edu/spcl/wise/Dreyfus/dreyfus.html.

Chemical Heritage Foundation Oral History Project. *Women in Science*. http://www.chemheritage.org/research/policy-center/oral-history-program/projects/women-in-science.aspx.

The Faces of African Americans in Science. https://webfiles.uci.edu/mcbrown/display/faces.html.

The ScienceMakers. http://www.thehistorymakers.com/biography/category_details.asp?sp=1&category=scienceMakers.

Just Garcia Hill, A Virtual Community for Minorities in Science. http://justgarciahill.org/index.php.

Jordan, Lynda, Yvonne Smith, and Michelle Pfeiffer. 1995. *Jewels in a Test Tube* (VCR available by interlibrary loan).

McLean, Lois, and Tessman, Richard, *Telling Our Stories: Women in Science.* McLean Media. 1996 & 1997 Program Guide and CD. http://www.storyline.com.

Sloan Career Cornerstone Center Career Planning Resources in Chemistry. .http://www.careercornerstone.org/chemistry/chemistry.htm.

University of Michigan. 1983. "Chemistry Careers for Women."*Women in Science Videotape Series*. Ann Arbor, MI: The School of Dentistry.

INDEX